Micaela Montano

Compost use and seasonal variation: quality assessment of soil

AF138594

Micaela Montano

Compost use and seasonal variation: quality assessment of soil

Ecotoxicological tools

Edizioni Accademiche Italiane

Impressum / Imprint

Bibliografische Information der Deutschen Nationalbibliothek: Die Deutsche Nationalbibliothek verzeichnet diese Publikation in der Deutschen Nationalbibliografie; detaillierte bibliografische Daten sind im Internet über http://dnb.d-nb.de abrufbar.

Bibliographic information published by the Deutsche Nationalbibliothek: The Deutsche Nationalbibliothek lists this publication in the Deutsche Nationalbibliografie; detailed bibliographic data are available in the Internet at http://dnb.d-nb.de.

Coverbild / Cover image: www.ingimage.com

Verlag / Publisher:
Edizioni Accademiche Italiane
ist ein Imprint der / is a trademark of
OmniScriptum GmbH & Co. KG
Heinrich-Böcking-Str. 6-8, 66121 Saarbrücken, Deutschland / Germany
Email: info@edizioni-ai.com

Herstellung: siehe letzte Seite /
Printed at: see last page
ISBN: 978-3-639-77621-8

*A **Gaetano***

- *Mi vuoi sposare?*

*To **Gaetano***

- *Will you marry me?*

INDEX

CHAPTER 1

CHAPTER 2

CHAPTER 3

CHAPTER 5

CHAPTER 6

INTRODUCTION (CHAPTER 1)

1.1 SOIL

Soil allows plants to grow and it is the key factor which controls water destiny in Earth environments. The soil is the system in which the recycling process of nature occurs, where vegetal and animal wastes are decomposed and transformed in their basic elements. Lastly, it is the habitat of many living beings, from small mammals to microscopic organisms. Although everyone knows it, soil is difficult to be defined. Without doubt, soil is not simply an abiotic environment for plants, but it is crawling with life and contains billions of bacteria, fungi and animals of different size. Biotic and abiotic interactions make the soil a living system. Soil scientists see the soil as a unit, or a tridimensional body, with its length, width and depth, and it changes perpetually in time and space (Smith and Smith, 2009).

It was in the late 1880s that the Russian Vasilij V. Dokuchaev, the father of pedology (arising from the Greek πέδον, *pedon* "soil" and λόγος, *logos* "knowledge"), gave dignity to soil as something with its own identity in the realm of natural objects. Dokuchaev, in fact, proposed a naturalistic concept of soil that does not deal with soil use. Essentially, he referred to the soil as a tridimensional entity located at the Earth's surface with morphology and unique physical, chemical and biological properties acquired by the interaction, through time, among living and dead organisms, rock, and climate on a given topographic position. In this view, soil was not considered as a weathered porous phase of bedrock from which plants take water and nutrients and to which they release their mortal remains (Certini and Ugolini, 2013).

In fact, today the soil is considered a unique and limited resource and is required for the production of food in order to maintain the quality of the environment (Jenny, 1980).

Although it varies in depth and properties, a soil is always formed, along its depth, by distinct layers, often having different composition and properties. These layers are defined soil horizons and the series of horizons from the surface downward is called the soil profile. The soil profile is given by: the upper horizon, called A horizon (surface soil); the next main horizon called B horizon, formed by the mineral components in which the organic compounds have been processed by decomposition into inorganic compounds (mineralization); the third horizon, C horizon, consisting of more or less unchanged parent material (Figure 1).

In a mature soil, the A horizon is usually divided into different layers that represent the successive stages of humification (Odum and Barrett, 2007). These layers are called (from the surface down) A-0 (litter), A-1 (humus) and A-2 (percolation area). Litter is the layer that covers the soil and can be considered a kind of ecological subsystem in which micro organisms (bacteria and fungi) with small soil invertebrates collaborate to decompose organic matter.

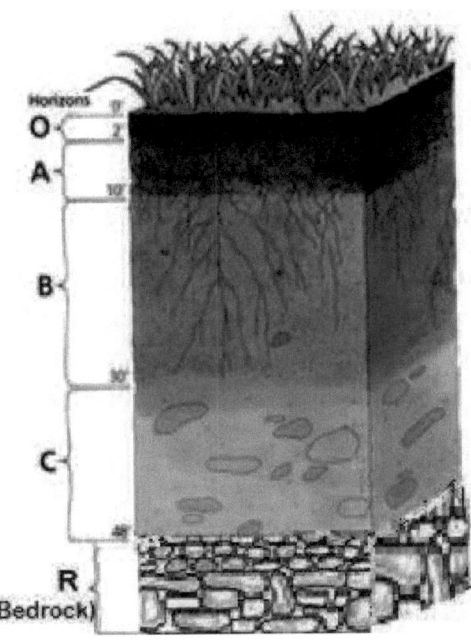

Figure 1. Soil profile: the upper horizon, called A horizon (surface soil); the next main horizon called B horizon, formed by the mineral components in which the organic compounds have been processed by decomposition into inorganic compounds (mineralization); the third horizon, C horizon, consisting of more or less unchanged parent material. *Image taken from Wikipedia.*

1.2 SOIL PROPERTIES

Soils have distinctive **physical characteristics**. These include color, texture, structure, water content and depth, all of which can vary considerably from one soil to another.

In addition, the main **chemical properties** of a soil are: the content of macronutrients and micronutrients, pH, ion exchange capacity, buffering capacity and organic matter content.

Finally, soils also have **biological properties**, like microbial biomass, respiration, total fungal biomass, metabolic quotient and coefficient of endogenous mineralization.

1.2.1 Physical Properties of Soil

Color is one of the most useful and easy to define features of a soil. It has little direct influence on the operation of a soil, but can be related to physical and chemical properties.

The texture of a soil is defined by the percentage (by weight) of the three main mineral components sand (S), silt (Si) and clay (C) in the soil (Figure 2). The ratio between the weight of these fractions and the total weight of the sample multiplied by 100 defines the percentage of particles present in the soil, based on their diameter size. A sieving procedure is carried out for assessing the larger particle sizes until the sand fraction (0.2 to 0.1 mm). For the determination of the finer fractions (fine sand, silt and clay), other methods are applied. The most famous ones are the method of sedimentation or densimetry and the method of smoothing. Both are based on the physical fact that a solid particle of definite shape immersed in water moves downward with a speed that depends on its size and its specific weight. Soil texture influences porosity, which plays a crucial role in enabling the movement of air and water in the soil and the penetration by plant roots. In an ideal soil, particles make up about 50% of the total volume; the remaining 50% being formed by pores. The volume occupied by the pores includes both the spaces between the soil particles and also the old tunnels earlier occupied by roots or dug by animals. Soils with a coarse texture have large pores, which facilitate infiltration, percolation and drainage of water. Till a certain point, the thinner is the soil texture, the smaller the pores in the soil will be, and then, the greater the availability of active surfaces for water contact and chemical activity will be. Heavy soils, with a very fine texture, such as clays, become compact easily if stepped on, are poorly ventilated and hardly penetrable by plant roots (Smith and Smith, 2009).

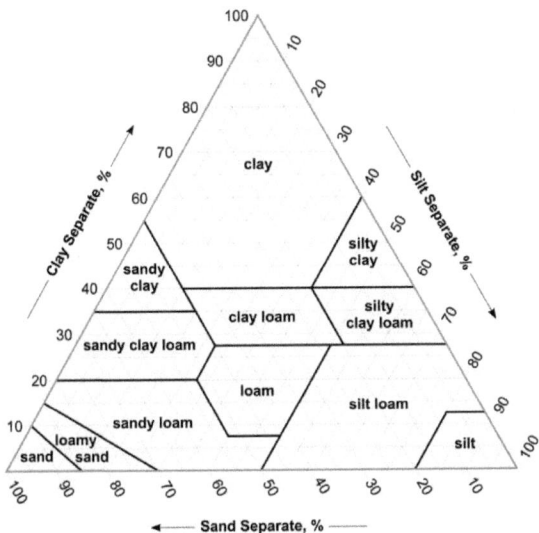

Figure 2. A diagram of soils texture which shows clay percentages (less than 0.002 mm), silt (0.002-0.05 mm) and sand (0.05-2.0 mm).

Image taken from USDA (United States Department of Agriculture).

Soil <u>structure</u> describes the way in which the primary particles, that is sand, silt and clay, interact and form aggregating particles. Soil structure influences some key factors in plant growth: ventilation, that is soil porosity; permeability and hydraulic conductivity; temperature and humidity of the soil; root growth; biological activity; leaching of bases and clay; resistance to soil erosion.

The <u>water</u> (moisture) <u>content</u> of soils is the ratio, expressed as a percentage, of the mass of "pore" or "free" water in a given mass of dry soil. The water content is an extremely important index used for establishing the relationship between the way a soil behaves and its properties. The consistency of a fine-grained soil largely depends on its water content. The water content is also used in expressing the phase relationships of air, water, and solids in a given volume of soil.

The underlined depth of the soil varies in different parts of an area based on the slope, the level of alteration from initial stage to intermediate or end stage, the parent materials and the vegetation. On flat terrain at the base of slopes and in floodplains soils tend to be deep. On the contrary, the soils on crests and on steep slopes have a low thickness and the parent rock is close to the surface. The soils that develop under prairies primary tend to be several meters deep, since a large part of the organic matter present in this type of vegetation is represented by the deep and fibrous root systems of herbaceous plants.

1.2.2 Chemical Properties of Soil

Nutrients represent that category of ions or molecules whose direct or indirect intake is essential for living organisms. *Macronutrients* are substances that are found more frequently in all major biological molecules. Oxygen, nitrogen, carbon and hydrogen are the predominant elements in such molecules, but also of great importance for life are sulfur, phosphorus, sodium, potassium, calcium, magnesium and chlorine in the form of the chloride ion. *Micronutrients* are required in extremely small quantities, and are found mostly within the molecules of enzymes or those having the role of "electron exchangers", such as cytochromes, chlorophylls, and carotenoids. The most important micronutrients are: iron, manganese, zinc, copper, cobalt, nickel, selenium, molybdenum, chromium, iodine as iodide and silicon. For the nutrients that are absorbed from the soil, it is necessary to consider their "availability": it is not enough, in fact, that they are just present in the soil, but it is necessary that they are in a physical-chemical state that makes them accessible to living organisms.

Since the solubility of nutrients is pH dependent, pH also is important for the activity of the microorganisms responsible for the decomposition of organic matter and most of the chemical transformations that occur in the soil. Sandy soils, in which water may more easily percolate, are theoretically more acidic than clayey soils (Bullini *et al.*, 1998).

The ion exchange capacity is important for soil fertility. An ion is a charged particle. The ions that carry a positive charge are called cations, while those that have a negative charge are anions. Elements and chemical compounds are present in the soil solution both as cations, such as calcium (Ca^{2+}), magnesium (Mg^{2+}) and ammonium (NH_4^+), and as anions, such as nitrate (NO^{3-}) and sulfate (SO_4^{2-}). The possibility that these ions have to bind to the surface of soil particles depends on the number of sites with positive or negative charge present in the soil. The total number of charged sites on the particles in a given volume of soil is mentioned the ion exchange capacity. In most soils of the temperate zone, cation exchange is the predominant process, because of the prevalence in the soil of particles with negative charge. Clay particles are the negatively charged constituents of soils. These negatively charged particles (clay) attract, hold and release positively charged nutrient particles (cations). Organic matter particles also have a negative charge to attract positively charged cations. Sand particles carry little or no charge and do not react. The cation exchange capacity, therefore, makes that the soil has the ability to retain positive ions on the surface of its organic and minerals components. Since the amount of cations accumulating on the soil particles always is in equilibrium with the concentration in the porewater, a decrease of concentration in the porewater will lead to a release of cations from soil particles.

Buffer capacity of soil is defined as a soil's ability to maintain a constant pH level during action on it by an acidifying or alkaline agent. A soil, considered a mixture of buffered systems, contains components, which have the ability to neutralize acids by binding H^+ ions as well as bases by the release of hydrogen ions (Federer and Hornbeck, 1985). The effectiveness of soil buffering systems depends on numerous physical, chemical, and biological properties of soils.

The soil organic matter (SOM) is the largest reservoir of terrestrial carbon (C). The soil contains 1500 billion tons of organic C, while in the atmosphere 720 billion tons of C is present in the form of carbon dioxide and only 560 billion tons of C is found in plant biomass. These forms have remained in a stable equilibrium until the advent of the industrial era, when the use of fuels and deforestation have led to a sharp decrease in plant biomass and soil organic matter, resulting in increased carbon dioxide levels in the atmosphere. The soil organic matter therefore, in addition to the extreme importance as a source of nutrients for plants, also has an important role as a carbon sink: "Soil organic matter is the fuel that runs the soil's engine" (Fisher, 1960). Hence the need to fully understand what is organic matter, and as it turns, what are the factors that influence the mineralization and accumulation and, finally, what are its functions in the soil. In soil organic matter we can find: simple substances such as amino acids and sugars; organic acids, such as mono and dicarboxylic acids; high molecular weight compounds such as polysaccharides, proteins, nucleic acids, lipids and lignins; humic substances such as humic and fulvic acids and humin (Figure 3).

Soil organic matter transformation processes, caused by the edaphic flora and chemical-physical processes, produce *humus*, which plays a pivotal role. *Humus* increases the water retention capacity of a soil, is a reservoir of nutrients and a substrate for plants, microorganisms and soil fauna (Pignatti *et al.*, 2004).

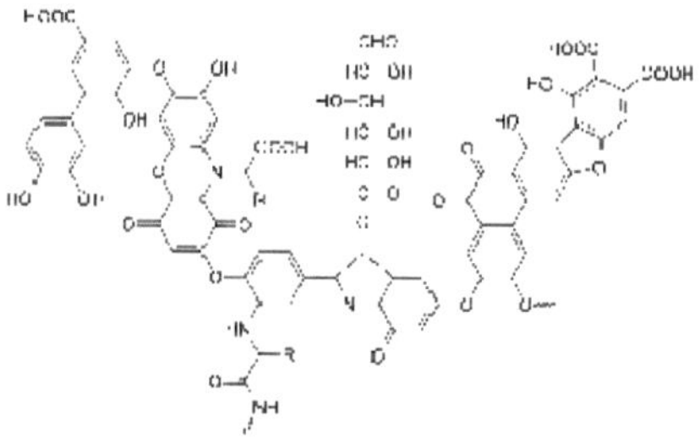

Figure 3. A molecule model of humic acid. *Image taken from Wikipedia.*

1.2.3 Biological Properties of Soil

The <u>microbial biomass</u> consists mostly of bacteria and fungi, which decompose crop residues and organic matter in soil. This process releases nutrients, such as nitrogen (N), into the soil that are available for plant uptake. Generally, up to 5% of the total organic carbon (C) and N in soil is in the microbial biomass. When microorganisms die, these nutrients are released in forms that can be taken up by plants. Microbial biomass is also an early indicator of changes in total soil organic C. Unlike total organic C, microbial biomass C responds quickly to management changes. Soil properties such as pH, clay, and the availability of organic carbon all influence the size of the microbial biomass.

Respiration is the biological activity of the entire soil biota (soil microbes, plant roots and micro and macro invertebrates, e. g. earthworms, arthropods and nematodes). The rates of soil CO_2 efflux vary by ecosystem (Reich and Schlesinger, 1992).

Total fungal biomass and also bacterial biomass are expressed in mg/g of dry weight.

The microbial metabolic quotient (respiration-to-biomass ratio) or qCO_2, and finally the coefficient of endogenous mineralization (CEM) are other microbial indices.

The importance of biological processes for the function of a soil ecosystem is unequivocal, and a large number of soil processes are, on a superficial level, dependent on biological activity. Microorganisms are essential for plant growth as they convert organic nitrogen, phosphorus, sulfur, and other nutrient elements into their inorganic state. They are also important for creating the soil environment where they themselves, soil animals, and plant roots are living. They contribute to both the quantitative and qualitative outcome of the growing plant.

Furthermore, in modern society, soil microorganisms are irreplaceable in the transformation of synthetic organic compounds and natural waste materials into organic forms that are environmentally acceptable (Tarradellas, Bitton and Rossel, 1996).

1.3COMPOST

The term "compost" derives from the Latin *compositum*, i. e. *mixed*: in fact, at the origin of the process of composting, there is almost always the mixing of different waste (Ciccotti *et al.*, 1988), which can derive from agricultural or from other sectors, that is household, urban, industrial waste (Figure 4).

The use of compost as a soil improver reduces the use of non-renewable resources used to produce chemical fertilizers, while reducing the amount of organic waste. As for all soil improvers, the use of the compost has the function of improving the quality of the soil, allowing to maintain in long-term fertility, soil structural state, the ability to absorb and release water and to retain nutrients in a form easily assimilated by plants, promoting the biological activity of the soil.

Compost can also prevent the colonization of pathogens thanks to mechanisms of antibiosis and to the competition of saprophytes contained in it. In fact composting is an effective means for reducing pathogen concentrations in a variety of organic materials, including manure, yard trimmings, and biosolids (sewage sludge). These materials, when "raw" or not composted, contain pathogens that may infect humans when they are first generated. The pathogens of interest include bacteria, protozoa, viruses, fungi, and helminths (parasitic worms). During the composting process, beneficial microbial populations build up while pathogen concentrations are considerably reduced. Therefore composting can be an effective means of reducing pathogens to acceptable levels in organic soil amendments. High temperatures associated with the composting process are responsible for killing pathogens.

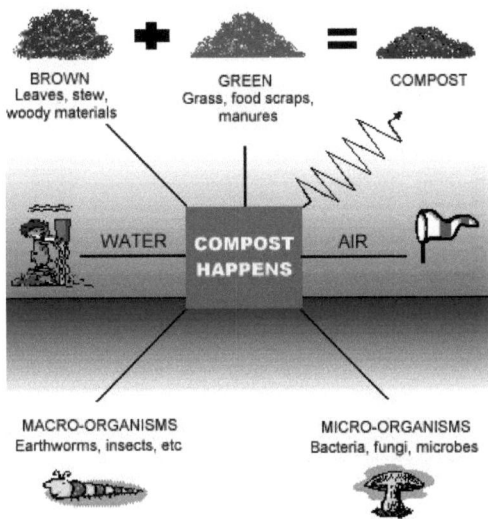

Figure 4. Composting is a mixture of decaying organic matter used to improve soil structure and provide nutrients. *Image taken from Website: www.yorkton.ca.*

Compost, being a product resulting from a conservative energetic process, has a high carbon content in its organic matter (Centemero and Caimi, 2001).

Because of its chemical characteristics, compost creates and maintains in the soil a suitable environment for the decomposition of organic substances.

As its own decomposition proceeds very slowly, compost acts as a reservoir of nourishment placed in the soil. Furthermore, it is believed that due to its high adsorption capacity, compost has the ability to slow the migration of contaminants in the soil.

In recent years a strong interest has developed in the potential role of compost in the fight against the greenhouse effect and its consequent climate change. Thus the biggest benefit of composting with respect to Global Climate Change comes from avoiding the production of methane. Good composting practices minimize greenhouse gas emissions. The use of compost provides numerous greenhouse gas benefits, both directly through carbon sequestration and indirectly through improved soil health, reduced soil loss, increased water infiltration and storage, and reduction in other inputs.

The organic fertilization continuous practices over time allow to maintain or increase the organic matter content in the soil, limiting the transfer of carbon, in the form of CO_2 to the atmosphere (Centemero and Zanardi, 2006). The European Commission's Directive on Soil Strategy strongly emphasizes the role of organic fertilization both to combat desertification and to promote the "sequestration" of carbon within the soil, contributing to the fight against climate change.

It should also be noted that Mediterranean Europe is included as an area of "desertification risk" according to the International Convention for the Fight against Desertification. A significant trend of the declining soil fertility is found in the countries of Southern Europe (Zdruli, Jones and Montanarella, 2004). This is for instance shown for Italy, where there have been many cases of soils, and not just in the South, but even in the northern plains, with nowadays less than 1% organic matter, which always had values above 2% (Van-Camp et al., 2004). In addition, strong salinization occurs with the intrusion of marine waters in the continental aquifers, increasing the level of salts in soils with obvious repercussions on agricultural yields. Also the process of erosion, which occurs where the passage of water causes the removal of soil, is affecting Italy for over a quarter of its territory. Where decreases in the organic matter content of soils can be recognized, negative effects can be found on crop production yields.

In summary, a soil that gradually loses its organic matter content will also face "decreases in its biological potential".

The use of compost as an organic amendment improves the permeability of the soil, prevents erosion and retains moisture. It may represent one of the answers to these problems. This also counts for the use of other natural soil improvers, like for example manure.

Compost, when added to the soil, becomes a growth factor for the plants and the source of production of new organic matter (Gallardo-Lara and Nogales, 1987). In addition, compost increases the biodiversity and activity of microbial populations in the soil, affecting the structure, nutrient cycling, and many other physical, chemical and biological processes (Albiach *et al.*, 2000). Indeed, the biological richness of the compost is an element of considerable importance, due to the bacterial colonies contained in it, capable of increasing the level of organic components of the soil (Scagliarini, 1999). Some effects of the use of compost on soil characteristics depend on the amount of organic carbon, organic and inorganic nitrogen, the degree of maturation of the compost, the content of heavy metals and the addition of mineral fertilizers (Crecchio, 2004).

The application of organic wastes to soil is a good management practice for reclaiming degraded soils, improving soil structure, increasing soil microbial biomass and sustaining soil microbial activity and thus the cycle of matter and macro and micro nutrient availability to plants (Caravaca *et al.*, 2002); at the same time it is a good ecological alternative to recycle such residues. However the occurrence of heavy metals and other contaminants in organic wastes represents a potential environmental hazard (Ihnat and Fernandes, 1996; Soumarè *et al.*, 2003); nevertheless the effects of organic wastes application on soil properties and plant yield depend on their components as well as on the amount of added organic material (Barzegar *et al.*, 2002).

Organic waste could be a source of soils recovery, but one has to pay attention to its advantages and disadvantages.

Advantages of organic composts: organic fertilizer is surely more advantageous than the chemical one in terms of respect and healthiness of the environment rather than of costs. In fact, not only does it prevent the soil from being excessively fertilized, but it also reinforces plants and composition of the soil, by avoiding the impoverishment of nutritive substances and the following desertification of farmed lands. Organic fertilizers can be cheaper than the chemical ones, especially when products easily found *in situ* are used.

<u>Disadvantages of organic fertilizers</u>: natural fertilizer has some disadvantages when used in farming on a large scale. Due to marketing requirements, it is often difficult to consider the use of organic fertilising, due to its long processing time. That is why, organic fertilizing is usually combined with the chemical one. Furthermore, the transportation of organic fertilizer would be more expensive than the chemical one in fertilizing on a large scale.

As interest in **Municipal Solid Waste (MSW)** composting increases, one of the concerns that must be addressed is the extent to which the low concentrations of heavy metals and metalloids (metal-like elements) present in MSW compost may adversely affect plant growth, soil organisms, water quality and animal and human health. Metals appear in the municipal solid waste stream from a variety of sources. Batteries, consumer electronics, ceramics, light bulbs, house dust and paint chips, lead foils such as wine bottle closures, used motor oils, plastics, and some inks and glass can all introduce metal contaminants into the solid waste stream. Composts made from the organic material in solid waste will inevitably contain these elements, although at low concentrations after most contaminants have been removed. Other trace elements (e. g., arsenic, cadmium, lead, and mercury) are of concern primarily because of their potential to harm soil organisms and animals and humans who may eat contaminated plants or soil.

Compost is increasingly used in order to restore the fertility of intensively farmed lands. Its usage as fertilizer has several advantages, mainly linked to the improvement of the chemical and physical components of farmed land, but it requires an accurate analysis of long-term risks. Therefore, it is important to evaluate the presence of polluting agents in compost and their possible toxic effects, to understand the influence of compost on polluting agents themselves and to monitor their possible transfer in food chains.

Adding a composted soil improver can spread toxic elements into the soil and, because of the bioaccumulation of heavy metals, could lead to a higher concentration in farmed lands (Iglesias-Jimenez and Alvarez, 1993) and, subsequently, transfer to man. The frequent addition of compost could cause the accumulation of metals and modify exchange processes, thus provoking a higher mobilization of less soluble and highly complexes heavy metals in the soil (Baldantoni *et al.*, 2010).

1.4ECOTOXICOLOGY

Ecotoxicology studies the toxic effects of chemical and physical agents on populations or communities within an ecosystem (Figure 5). Ecotoxicology, then, identifies the different types of transport of these agents and their interaction with the environment (Butler, 1978). Therefore, ecotoxicology studies the mechanisms of action of pollutants, assesses the biological damage in one or more species, but in addition it also assesses effects at the ecosystem level by integrating the effects of stress factors across all levels of biological organization from the molecular to communities and ecosystems (Walker, Sibly, Hopkin and Peakall, 2012). Ecotoxicological surveys also take account of the interactions between compounds. These compounds can give rise to phenomena of synergy or antagonism. Finally, ecotoxicological surveys also takes account of all the substances that are not characterized as toxic by definition, but may produce trophic imbalances. Then these substances alter the composition of ecosystems in a more or less permanent way (APAT, 2006). Ecotoxicological analysis are carried out through the use of ecotoxicological bioassays, that is just one way of approaching the risk of soils or sediments. An ecotoxicological assay is a test, usually performed in the laboratory, which uses a biological system to assess the toxic effects induced by exposure, for a given time, to the toxic substance or environmental matrix under consideration. Endpoints in such tests are alteration or impairment of one or more functions such as survival, growth, reproduction, motility, photosynthesis, or behavior (Maffiotti *et al.*, 1997). Because the effect is manifested it is necessary that the substance not only enters the body of the test organisms, but also comes in contact with the cell structures (possibly with a specific target site) in an amount and for a time sufficient to induce biological damage. A toxicity test is based on the principle that exposing a living organism to a toxic agent will lead to a response that is a direct function of the dose level.

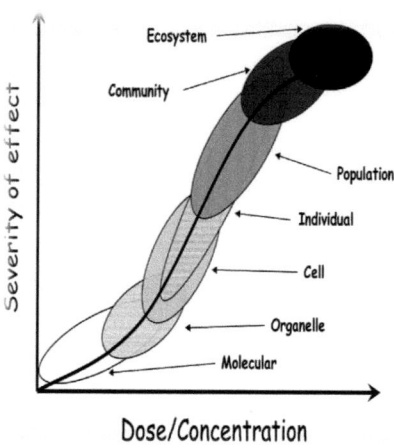

Figure 5. The relationship between biological organization, concentration and effects in ecotoxicology. *Image taken from the lecture in the course Ecotoxicology.*

The toxicity of an agent (or mixture) generally first results in effects at the biochemical and molecular level (change in enzymatic activity, DNA alterations) and only subsequently at the level of organelles, tissues, and finally at the level of the individual and the population.

The type of effect that is measured in toxicological assays is typically referred to as the "endpoint". The endpoint can be survival, reproduction or growth or be related to metabolic or physiological parameters of the organism ("biomarkers") depending on the type of assay used.

Ecotoxicological assays can be classified according to the duration of exposure and the life span of the organism (Wright and Welbourn, 2002):

• Acute tests. Short-term, detect adverse effects that occur in a short period of time after administration of a single dose of the substance.

• Sub-chronic tests. Show effects due to exposure to a substance for a period of less than or equal to one-tenth of the life of the organism.

• Chronic tests. Estimate the effects that occur as a result of exposures for a longer time, which often coincides with more than half of the duration of life span of the test organism.

Toxicity tests conducted directly on the solid matrix assess the interactions between the soil and the contaminant that determine the bioavailability of toxic substances present in the soil. Tests conducted on the solid matrix therefore are more realistic because they use the matrix *in toto* (APAT, 2004).

Batteries of tests on the soil include tests with plants (Chung *et al.*, 2007; Hamdi *et al.*, 2007; Hubalek *et al.*, 2007; Leitgib *et al.*, 2007) and terrestrial invertebrates such as nematodes, earthworms, and springtails (Nahmani *et al.*, 2007; Crouau and Pinelli, 2008; Roh *et al.*, 2009). When growing on a contaminated soil, plants can accumulate toxic substances (e.g. metals) in their tissues, potentially threatening the health of animals and humans. Plants also have the property, through the production of exudates from the roots, to affect the soil environment, for instance by acidifying the soil and modifying the mobility of metals or other contaminants. To study the effects of soil contaminants on plants is very important because of their ecological role as primary producers. The endpoints normally used for plant tests are germination and root elongation, but it is also possible to perform studies on biomass production and bioaccumulation of metals and other contaminants (An, 2004). The species of soil invertebrates commonly used in ecotoxicity testing are selected according to their sensitivity to contaminants, ease of breeding in the laboratory, high rate of reproduction and speed of response to exposure (Van Gestel, 2012).

Ergo, ecotoxicological tests are useful to evaluate the soil remediation effectiveness.

AIM OF THE RESEARCH (CHAPTER 2)

My AIM is to assess the influence of the application of COMPOST "on soil quality". Soil quality is defined in terms of both concentrations of metals and the bioavailability of the metals as determined using ecotoxicological assays.

To this aim, the following physico-chemical parameters were measured: pH, Water Holding Capacity (WHC), Water Content (WC), Organic Matter (OM) content, content of carbon (C) and nitrogen (N), total concentrations of metals (Cd, Cr, Cu, Ni, Pb), available metal concentrations.

These biological parameters were assessed: microbial biomass (Cmic), respiration, Total Fungal Biomass (TFB), metabolic quotient (qCO2), Coefficient of Endogenous Mineralization (CEM).

Ecotoxicological tests showed: germination and root extension in Lepidium sativum *and in* Sorghum saccharatum, *and survival and growth of* Heterocypris incongruens.

Also metal bioavailability was measured with the earthworm Eisenia andrei.

My research question is:

"What is the possible impact of organic amendment on substrates quality?"

The subject of research is the "quality assessment of soil" using ecotoxicological tools.

2. MATERIALS AND METHODS (CHAPTER 3)

2.1 Sampling

Soil sampling was carried out in October 2013 and in March 2014 in 13 mesocosms of the wreckage of a mine no longer used. Mesocosms were pots (1m diameter, 60 cm height) containing limestone debris covered by the organic material. An additional sampling was carried out in September 2014 for an in-depth ecotoxicological analysis.

The following abbreviations have been used: F refers to Fillirea (*Phillyrea angustifolia L.*), A to Alloro, Bay tree, (*Laurus nobilis L.*) and L to Leccio, Holly oak, (*Quercus ilex L.*), all of them green coverages (sclerophyllous shrubs).

P stands for Pollina (Poultry manure) which, along with COMPOST, is an **organic compost**. Pollina is an organic manure from the industrial recycling of avicultural breedings.

The following nomenclature has been used: αPFA, βPFA, αFA, βFA, αPFL, βPFL, αFL, βFL, αPFAL, βPFAL, αFAL, βFAL, COMPOST (when α and β are the repetitions of the field).

For each mesocosm, at each sampling time, three samples of substrate were collected by a sampler (5 cm Ø) from the upper 10 cm layer and mixed into a representative composite sample.

After collection, the soil samples were sieved (2 mm) for subsequent chemical-physical and ecotoxicological analysis and stored in the dark at 4 °C.

2.2 Physicochemical characterization of substrates

After sampling, the soils were used to determine: pH, maximum water holding capacity, water content, organic matter content, content of carbon and nitrogen, and the total and available concentration of five metals (Cd, Cr, Cu, Ni, Pb).

2.2.1 pH

The pH of the soils was determined using a potentiometric method.

For that purpose, 25 mL of deionized water were added to 10 g of fresh soil, weighed in an Erlenmeyer flask.

The flasks were shaken oscillating surface for 20 minutes and then left for another 10 minutes to allow settling of the soil particles.

Subsequently the pH of the supernatant was measured using a pH meter.

2.2.2 Water Holding Capacity (WHC)

The water holding capacity was determined by the gravimetric method on cool soil.

For the measurements, plastic cylinders with a known weight, open at the top and with holes at the bottom-covered internally by blotting paper-were used.

The cylinders were filled with soil. The weighed soil samples were partially immersed in deionized water, in order to obtain the complete soaking of the sample.

The weight at saturation was determined and the samples were dried in an oven at 105 °C until reaching constant weight.

The water holding capacity was calculated with the following formula:

$$WHC\ (\%) = ((\text{net weight at saturation - net weight of dry soil}/ \text{net weight of dry soil})* 100$$

The determination of the WHC was performed on three replicates of each sample.

2.2.3 Water Content (WC)

The determination of water content was performed by weighing 5 g of fresh sample, dry it-in an oven at 105 °C until reaching constant weight and weighing it again.

The water content of the samples was expressed as percentage of the dry weight of the sample and was calculated with the following formula:

$$WC\ (\%) = ((\text{net weight of fresh soil - net weight of dry soil}/ \text{net weight of dry soil})* 100$$

There were three replicates for each sample.

2.2.4 Organic Matter (OM) content

The organic matter content of the soil samples was determined by incineration.

Therefore 1 g of sample was weighed, dried in stove at 105 °C for 48 hours in calibrated porcelain capsules, and subsequently placed in a muffle furnace at 550 °C for 2 hours.

The organic matter content was expressed as the percentage of the dry weight of the sample and was calculated with the following formula:

$$OM\ (\%) = ((\text{net weight of dry soil - net weight of burned soil}) / \text{net weight of dry soil}) * 100$$

There were three replicates for each sample.

2.2.5 Content of Carbon and Nitrogen

The determination of the content of carbon and nitrogen was carried out by elemental analyzer (Elemental Analyzer, Flash EA 112 Series).

Therefore 10 mg of dried and pulverized soil were weighed into tin capsules using an analytical microbalance.

The capsules were closed and placed in the sampler of the elemental analyzer.

Carbon and total nitrogen contents, stated as percentages on the dry weight sample, are provided by the instrument and are based on the calibration curve (ER) created using 4 standard holly oak leaves of growing weight (C= 49.81% and N= 1.855%), certified by Carlo Erba Instruments (Milan).

All samples were analyzed in three replicates.

2.2.6 Total concentration of metals

To determine the total concentrations of metals, the soils were dried in an oven at 105 °C and pulverized.

Next 4 mL of HNO_3 (65%) and 2 mL HF (50%) were added to 250 mg of dry soil sample, weighed into Teflon digestion tubes. For the blank the same quantities of the two acids without the addition of soil were used.

The tubes were hermetically closed and placed in a Milestone microwave oven (Digestor/Dryng Module mls 1200) with the following program for digestion:

1. 250 watt for 5 minutes;

2. 400 watt for 5 minutes;

3. 0 watt for 2 minutes;

4. 500 watt for 5 minutes;

5. 0 watt for 2 minutes;

6. 400 watt for 5 minutes;

7. 0 watt for 2 minutes;

8. 400 watt for 5 minutes;

9. 0 watt for 2 minutes;

10. 400 watt for 5 minutes.

After cooling in excess water at room temperature, the tubes were opened and the digestion solution was transferred to volumetric flasks and diluted with deionized water to a volume of 50 mL.

The samples were analyzed by atomic absorption spectrometry (Spectra AA220 FS, Varian).

The concentrations of Cd, Cr, Cu, Ni and Pb were determined by atomization using graphite furnace, starting from a calibration curve constructed using solutions with known and increasing concentrations of each considered element.

The total concentrations of Cd, Cr, Cu, Ni and Pb in soils were expressed in μg * g^{-1} dry weight of the sample.

Analyses were conducted in three replicates for each sample. Accuracy was checked by concurrent analysis of standard reference materials (light sandy soil) from the Community Bureau of Reference of the Commission of the European Communities (BCR No. 142R).

2.2.7 Available metal concentrations

For soils with a pH > 6.5 the extraction was performed by the method reported by Lindsay and Norwell (1978).

This method involves the extraction of metals with a solution made of 1.97 g of dietilentriamminopentacetico acid (DTPA), 1.47 g of hydrated calcium chloride and

14.92 g of triethanolamine, brought to a pH value of 7.3 ± 0.05 with 1N HCl, and then adjusted to a volume of 1000 mL with deionized water. Of this solution, 50 mL was added to 25 g of the soil sample previously dried in an oven at 105 °C.

The samples were stirred on a oscillating plane for 120 minutes, and filtered over a filter paper.

The extracts obtained were analyzed using the atomic absorption spectrometer (Varian Spectra AA 220 FS) with the same procedure as for the total content of metals.

2.3 Biological characterization of substrates

After sampling the soils were analysed for: Microbial Biomass (Cmic), Respiration, Total Fungal Biomass (TFB), Metabolic Quotient (qCO_2), and Coefficient of Endogenous Mineralization (CEM).

2.3.1 Microbial Biomass and Respiration

An estimate of the biomass and activity of the microbial community of the soil was made through the measurement of CO_2 produced by the oxidation of organic matter, which is soil respiration. Sieved soil was used to exclude the contribution of roots and macrofauna respiration.

Microbial biomass was evaluated using the method of substrate-induced respiration (SIR) according to Anderson and Domsch (1978).

To a quantity of fresh soil, equivalent to 5 grams of soil dry weight, 3 mL of a glucose solution at a concentration of 30 mg mL^{-1} were added.

The samples were incubated in jars placed in an airtight glass beaker containing approximately 15 mL of water (in order to ensure a moist environment) and 10 ml of a 0.1 N solution of NaOH. The jars were incubated for 5 days in the dark at 25 °C. The solution of NaOH acts as a "trap" for the CO_2 released from the soil. Adding an excess of D-glucose, easily usable by living species, it is possible to estimate the evolution of CO_2 and thus the activity of the microorganisms present in the soil samples under investigation.

The evolution of CO_2 is proportional to the amount of the microorganisms in the sample. In addition to estimating respiration, the samples of soil were incubated in the same conditions and for the same period, as for the determination of the microbial biomass with glucose, after adding only 3 ml of distilled water only. Doing so it is possible to estimate the basal respiration of soil. Together with the jars containing the samples of litter and soil with H_2O or with glucose addition, three blanks were incubated, i.e., jars containing only NaOH and the beaker with distilled water, to subtract the amount of CO_2 already present in the jar due to the activities of soil microorganisms. After the incubation, a double titration with two indicators, phenolphthalein and methyl orange, and a titrant solution of HCl (0.05 N), was performed. During incubation, the sodium hydroxide reacts with the CO_2 contained in the jar and produced by soil organisms, thus obtaining Na_2CO_3, according to the following reactions:

1) $2NaOH + CO_2 \rightarrow Na_2CO_3 + H_2O$

During the first titration with hydrochloric acid, which uses the phenolphthalein indicator, neutralizes the excess NaOH, which has not reacted with CO_2, according to the reaction:

2) $NaOH + HCl \rightarrow NaCl + H_2O$

Subsequently, the acid reacts with the sodium carbonate obtained from reaction (1):

3) $Na_2CO_3 + HCl \rightarrow NaCl + NaHCO_3$

When all the carbonate has been converted into bicarbonate, phenolphthalein turns from purple to colorless.

In the second titration orange methyl is used as indicator, which changes from yellow to orange when all the CO_2 is liberated from sodium bicarbonate, according to the following reaction:

4) $NaHCO_3 + HCl \rightarrow NaCl + CO_2 + H_2O$

The amount of carbon as CO_2 evolved from each sample is obtained using the following formula:

5) mg $CO_2 = 2* (a - b)* 1.1$

where:

a = ml of HCl used in the second titration of the sample

b = ml of HCl used in the second titration of the blank, which indicates the CO_2 present in the atmosphere of the jars, regardless of the soil.

The stoichiometry of the reactions shows that the evolution of one mole of CO_2 from sodium carbonate (reactions 3 and 4) requires two moles of HCl: one mole to convert the carbonate to bicarbonate (reaction 3) and a mole to release the CO_2 from bicarbonate (reaction 4).

Since the milliliters of HCl taken into account in the formula are only those of the second titration (reaction 4), these should be multiplied by 2. Also, knowing that each ml of 0.05 N HCl used in the titration equals 1.1 mg of CO_2, the milliliters of HCl used in the second titration should be multiplied by 1.1. The mg of CO_2 evolved are reported per gram dry weight of litter or soil and per time unit. In addition, the microbial biomass or microbial carbon was estimated using the following equation:

Microbial biomass = mg C-CO_2 g dry soil^{-1} h^{-1} x Kc (evolved after the addition of glucose), where Kc is the fraction of biomass carbon mineralized to CO_2 during 5 days of incubation (Jenkinson and Ladd 1981).

2.3.2 Total Fungal Biomass

The preparation of glass slides for this analysis started from a suspension of the samples of soil (0.5 grams) prepared with 50 mL of phosphate buffer (60 mM, pH 7.5), homogenized for 3 minutes at 6000 RPM; therefore 0.5 mL of suspension with soil and filtered under vacuum (with a filter of 25 mm diameter and a 0.45 μm of porosity), was collected.

The filters were stained with a solution of aniline blue that is able to bind chitin, which is a cell wall component of the fungal mycelium. The dye does not have the ability to discriminate between the hyphae metabolically active, inactive, or death and not yet decomposed. The filters (colored and dried) were built on a glass slide. The determination was carried out by counting the intersections of hyphae (Olson 1950), with the meshes of a net placed on the eyepiece, using a 40x magnification, and moving the view field horizontally and vertically up to 20 times. By the intersections between the hyphae and the mesh known dimension we obtained the length of the fungal mycelium.

The length of the hyphae was expressed in mm/g and multiplied by the mean section of hypha ($9,3* 10^{-6}$) to get the volume in $mm^{-3}g$ and then expressed on a dry weight basis, knowing the average density of the hyphae (1.1 g ml^{-1}) and knowing that the dry weight of one hypha represents 15% of the fresh weight (Berg & Söderström 1979). The fungal biomass was expressed as mg/g dry weight.

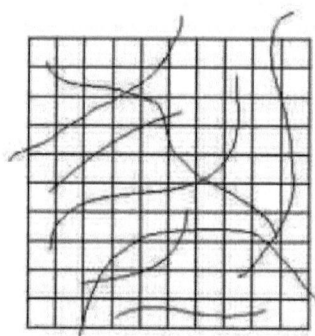

Figure 6. The screen built on the eyepiece of the microscope light and fluorescence assesses the intersection of fungal hyphae.

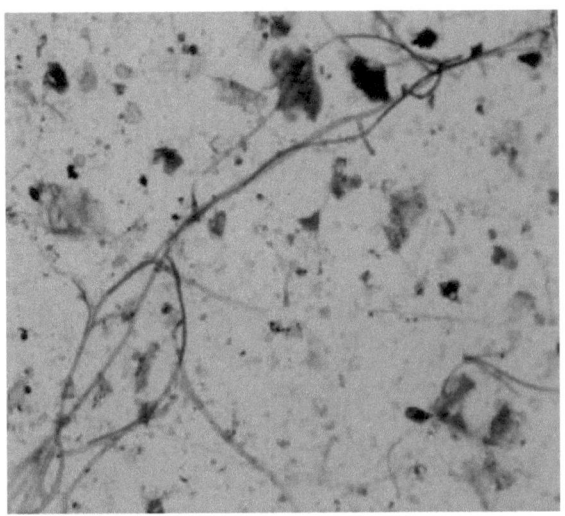

Figure 7. Fungal hyphae stained with blue aniline observed by the microscope light.

2.3.3 Metabolic Quotient (qCO₂) and Coefficient of Endogenous Mineralization (CEM)

In order to know the activity of the microflora present in the soil the metabolic quotient (qCO_2) was calculated. It represents the amount of carbon evolved as CO_2 per unit of microbial carbon (Anderson & Domsch 1993), and takes into account the duration of incubation:

$$qCO_2 = mg\ C\text{-}CO_2\ mg\ C\ mic^{-1}\ g^{-1}$$

The qCO_2 may be a useful index to identify the onset of stressful conditions for the edaphic microflora and also to differentiate soils at different stages of maturity. In addition to the assessment of the mineralization rate of organic C, the coefficient of endogenous mineralization (CEM) was calculated. It represents the ratio of carbon evolved as CO_2 during the incubation period and the content of soil organic carbon:

$$CEM = mg\ C\text{-}CO_2\ g\ Corg^{-1}\ g^{-1}$$

3.4 Ecotoxicological analysis

The ecotoxicological analysis were conducted on the soil as such.

The choice of this analysis is mainly based on several factors: the availability of standardized organisms in farming, the simplicity of the tests and the sensitivity of organisms.

3.4.1 Toxicity tests on substrates

The selected and applied ecotoxicological analyses of the sampled soils include:

✓ Acute and chronic phytotoxicity test with a monocot (*Sorghum saccharatum*) and a dicotyledon (*Lepidium sativum*);

✓ Acute and chronic test with Ostracod *Heterocypris incongruens*.

The tests were conducted on soil samples collected in both seasons (Fall and Spring).

3.4.2 Acute and chronic phytotoxicity test with Sorghum saccharatum *and* Lepidium sativum

Phytotoxicity assays offer the possibility to highlight the cumulative and/or synergistic effect of multiple contaminants.

This test evaluates the effects on germination and on radical growth (EPA, 1996) of a monocot *Sorghum saccharatum* (Sorghum) and a dicotyledon *Lepidium sativum* (cress).

The negative control was performed with an artificial soil (OECD, 1984), consisting of quartz sand (70%), kaolinitic clay (20%), peat moss (10%).

In Petri dishes (Ø 10 cm) a quantity of fresh soil, equivalent to 10 g dry weight, was weighed and covered with a filter paper (Ø 9 cm).

The amount of deionized water required for the saturation plus another 5 ml of water was added to the soil.

Ten seeds were placed on the surface of the filter in each dish.

The dishes were incubated for 72 hours in the dark at 25 ± 2 °C, in closed polyethylene bags to minimize water evaporation.

At the end of the test, the number of germinated seeds (root length ≥ 1 mm) was counted for each replication. The root tips were measured on graph paper.

To express simultaneously the two endpoints a germination index (GI) was calculated using the following formula:

GI = L* n

where L is the average length of the roots and n is the average number of germinated seeds.

The results were expressed as percentage of effect compared to the control.

3.4.3 Acute and chronic test with Ostracod Heterocypris incongruens

The ostracods are benthic organisms mainly living at the interface between water and sediment and feeding on solid particles. They are particularly sensitive to the toxicity released from the substrate.

The species *Heterocypris incongruens* is considered particularly suitable for toxicity testing of soil pollutants because it is sensitive to both organic and inorganic pollutants.

The assay with *Heterocypris incongruens* on a solid matrix has a duration of 6 days and evaluates two different endpoints: mortality (acute effect) and growth inhibition (chronic effect) compared to a control.

The assay was performed using newborn individuals-hatched from resting eggs (cysts) stored in the dark at 4 °C.

It used the artificial soil according to OECD (1984) as a control.

Before the execution of the test, *Heterocypris incongruens* cysts (kit obtained from a commercial supplier) were hatched in a Petri dish (diameter 5 cm) with 10 ml of Freshwater Standard (SF) for 52 h continuous light (3000-4000 lux) at a temperature of $25 \pm 2 ° C$.

The composition of the Standard Freshwater is the following:

✓ $MgSO_4$ 60 mg l^{-1};

✓ $NaHCO_3$ 96 mg l^{-1};

✓ KCl 4 mg l^{-1};

✓ $CaSO_4$ $2H_2O$ 60 mg l^{-1}.

The SF was stored in the dark at 4 °C before being used and ventilated and warmed to room temperature.

After 48 hours of the onset of hatching of the cysts, we pre-fed using the alga *Spirulina sp.* continuing the incubation for 4 more hours.

The test was performed in a six-well plate (Ø 4 cm) and neonates with a body length initial between 200 and 250 μm were used.

The procedure of Chial and Persoone (2003) was modified to add a quantity of fresh soil equivalent to 400 mg of dry soil in 4 ml of Standard Freshwater to each well and an algal suspension with a final concentration of $1,5* 10^7$ cells mL^{-1}. The suspension was stirred and decanted for 20 minutes.

Next 10 ostracods were transferred into each well, the plates were sealed with parafilm and lid and incubated in the dark at 25 ± 2 °C for 6 days.

At the end of the test the survivors were transferred to a multi-well plate for counting and for measurement of body length.

Stereomicroscope measurements (Zeiss Stemi 2000-C) were carried out by using a graduated scale of 50 μm. The ostracods were immobilized with a drop of fixative solution (Lugols).

The results were expressed as percentage of mortality (acute effect) and as a percentage of growth inhibition (chronic effect) compared to the control.

3.5 Assessing metal bioavailability with the earthworm Eisenia andrei

After sampling, soils were left to dry for about a week in order to carry out further ecotoxicological analysis at the Department of Ecological Science at the VU University Amsterdam.

Earthworms (*Eisenia andrei*) were used as test organisms to assess bioavailability of elements in the substrates samples. Before placing earthworms in the soil samples, they were taken from the culture and put in the control Lufa 2.2 soil to acclimatize for about 24 hours.

Subsequently, 200 g of each soil sample was moistened to 50% of its water holding capacity. Next, 1.4 g of *dried horse dung (food)* was added. After mixing, 100 mL jars were filled with approx. 45 g portions of each test soil.

Earthworms were removed from soil Lufa 2.2, washed, blotted dry on filter paper and randomly divided over the jars containing the soil samples, one earthworm per jar. Jars were placed in a room at a temperature of 20 °C (75% humidity, 18 h. light/ 6 h. dark). Five replicates were used for each soil sample. Exposures lasted for 21 days.

To evaluate possible effects on survival or body weight change, the jars were opened, the earthworms were collected from the soil by hand sorting, washed, blotted dry, weighted and placed in Petri dishes. The earthworms were incubated for 24 hours at 20 °C to empty their guts. Then they were placed in a freezer (-20 °C) for storage.

Earthworms were freeze dried, weighed and digested *by means of the bomb method*. For that purpose, individual dry worms were place in Teflon bombs, 2mL of a *destruction mixture* (HNO_3: HCl = 4: 1) was added and the bombs were tightly closed. *Destruction* lasted for 7 hours (*the digestion is at 140 °C*). Then 8mL of DEMI WATER were added dilute the digestion solution to 10 mL.

Afterwards, concentrations of 3 metals (**Cd, Ni, Pb**) were measured in the digests by flame atomic absorption spectrometry, using a PERKIN ELMER AANALYST 100. Finally, Cd, Ni and Pb concentrations in the earthworms were calculated using the following formula:

[(Measurement (mg/L) - blank)*10 mL]/Earthworms Weight (mg).

3. RESULTS AND DISCUSSION (CHAPTER 4)

3.1 PHYSICO-CHEMICAL CHARACTERIZATION OF MESOCOSMS

The substrates tested were characterized for pH, organic matter and moisture content (water content in the soil at the time of sampling) (Table 1 and Table 2).

Table 1. Average values (standard error) of pH, Organic Matter content and moisture content (% d. w.) of substrates sampled in October 2013.

October 2013	pH	O.M. (% d.w.)	W.C. (% d.w.)
α PFA	7.84	7.15 (1.68)	44.0 (0.250)
β PFA	7.87	10.2 (2.41)	48.2 (0.380)
α FA	7.90	14.2 (2.54)	53.7 (0.190)
β FA	7.94	14.0 (0.763)	47.6 (0.870)
α PFL	8.07	8.21 (0.556)	46.6 (0.120)
β PFL	7.92	13.7 (2.21)	51.3 (0.680)
α FL	8.04	13.3 (2.74)	40.4 (0.050)
β FL	7.86	13.3 (1.60)	49.9 (0.430)
α PFAL	7.98	8.75 (0.355)	45.3 (0.220)
β PFAL	7.91	9.51 (1.18)	40.6 (0.260)
α FAL	7.97	8.36 (0.655)	34.1 (0.380)
β FAL	7.83	15.2 (1.32)	51.5 (0.210)
COMPOST	7.91	20.6 (1.20)	69.0 (0.280)

Table 2. Average values (standard error) of pH, Organic Matter content and moisture content (% d. w.) of substrates sampled in March 2014.

March 2014	pH	O.M. (% d.w.)	W.C. (% d.w.)
α PFA	8.01	11.5 (1.79)	42.6 (0.645)
β PFA	8.04	11.5 (1.53)	38.0 (0.109)
α FA	8.19	6.47 (4.10)	31.1 (0.242)
β FA	8.05	19.7 (3.46)	55.9 (0.645)
α PFL	8.30	6.79 (8.49)	28.4 (0.590)
β PFL	8.14	26.5 (7.22)	66.2 (0.555)
α FL	8.12	15.2 (1.79)	45.3 (0.986)
β FL	8.12	33.6 (7.05)	53.0 (0.719)
α PFAL	8.11	21.5 (2.50)	57.0 (0.295)
β PFAL	8.07	27.0 (3.29)	50.0 (0.300)
α FAL	8.18	18.1 (9.05)	41.5 (0.093)
β FAL	8.13	14.5 (4.10)	39.0 (0.656)
COMPOST	8.11	22.7 (1.80)	65.9 (0.477)

In the graphs below, the results of treatments with FA (*Fillirea* and *Alloro*), FL (*Fillirea* and *Leccio*) and FAL (*Fillirea*, *Alloro* and *Leccio*) are combined and indicated by the abbreviation CP (compost with plants), while those for PFA, PFL and PFAL are indicated by CPp (compost with plants and poultry manure). The COMPOST-only treatment is indicated with C. The different colors show the differences between the two different sampling times (October 2013 and March 2014).

The addition of compost with poultry manure and plants (CPp) gave a higher water holding capacity (Figure 8) as compared to compost (C) and to compost plus plants (CP), although no significant difference occurred (ANOVA test). The addition of poultry manure to compost enhanced water holding capacity in agreement with the results of Eck and Stewart (1995).

The pH of the substrates sampled in autumn was around neutrality, while the pH of the substrates sampled in spring was slightly basic (around 8). Seasonality in this case seems to have led to a shift of pH from neutrality to near sub-alkalinity showing that the pH of a soil may vary throughout the year. In fact it depends on the conditions of temperature and humidity (increases and decreases in the winter season and in the summer season, respectively) (Figure 9). Statistical analysis of results showed that in all cases, pH was significantly lower in 2013 than in 2014 ($p<0,01$).

At both samplings the compost had the highest organic matter content and also the highest moisture content (Figure 10 and Figure 11), although the differences were significant only for organic matter content. Statistical analysis of results showed that Organic Matter content was significantly higher ($p<0,05$) in CP 2013 *versus* CPp 2013. The addition of poultry manure to compost did not increase organic matter content, which is in disagreement with the results of Eck and Stewart (1995).

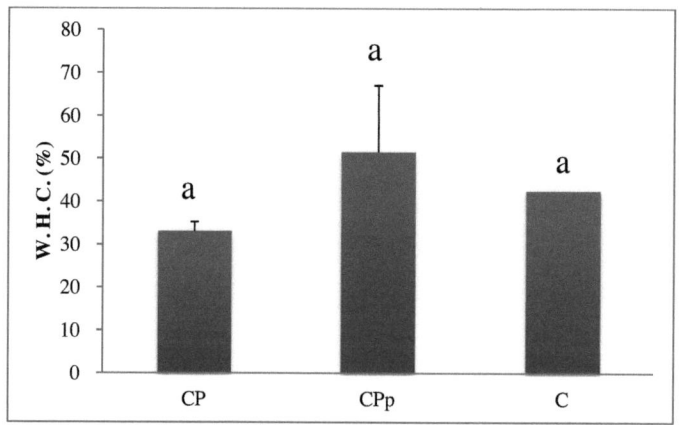

Figure 8. Average (± standard deviation) Water Holding Capacity (%) of substrates with different treatments sampled in autumn 2013 and spring 2014.
CP = Compost + Plants (n = 3); CPp = Compost + Plants + poultry manure (n = 3);
C = Compost (n = 1).

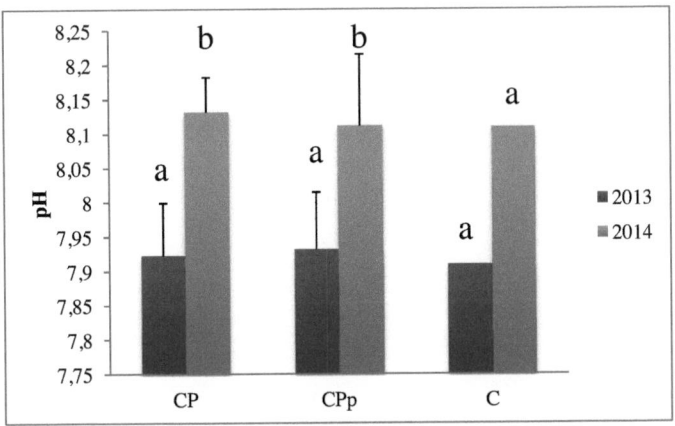

Figure 9. Average (± standard deviation) pH of substrates with different treatments sampled in autumn 2013 and spring 2014.
CP = Compost + Plants (n = 6); CPp = Compost + Plants + poultry manure (n = 6); C = Compost (n = 1).

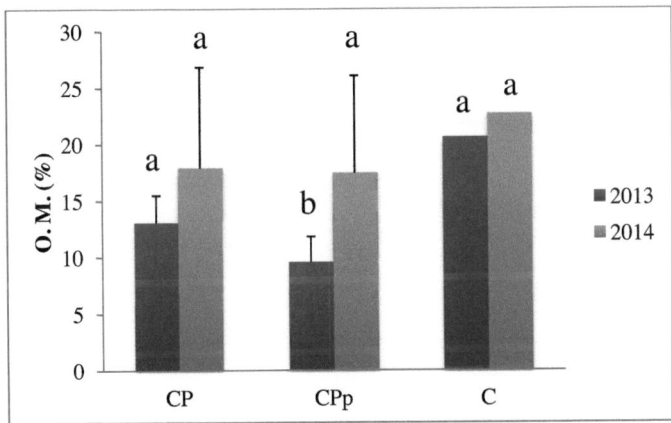

Figure 10. Average (± standard deviation) Organic Matter content (%) of substrates with different treatments sampled in autumn 2013 and spring 2014.
CP = Compost + Plants (n = 6); CPp = Compost + Plants + poultry manure (n = 6); C = Compost (n = 1).

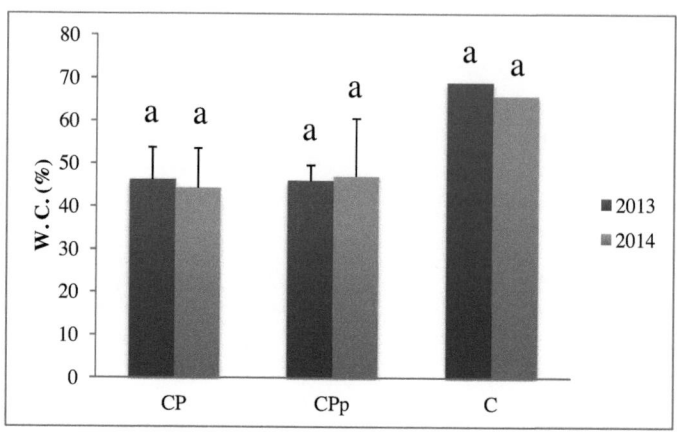

Figure 11. Average (± standard deviation) moisture content (%) of substrates with different treatments sampled in autumn 2013 and spring 2014.
CP = Compost + Plants (n = 6); CPp = Compost + Plants + poultry manure (n = 6);
C = Compost (n = 1).

The total contents of C and N were measured on dried and finely pulverized soils (Figure 12 and Figure 13), and followed the trend of the organic matter content, although the differences were significant (p<0,001) only for total content of N. They both tended to decrease due to microbial activity of the soil (Kaiser and Zech, 2000; van Hees *et al.*, 2003).

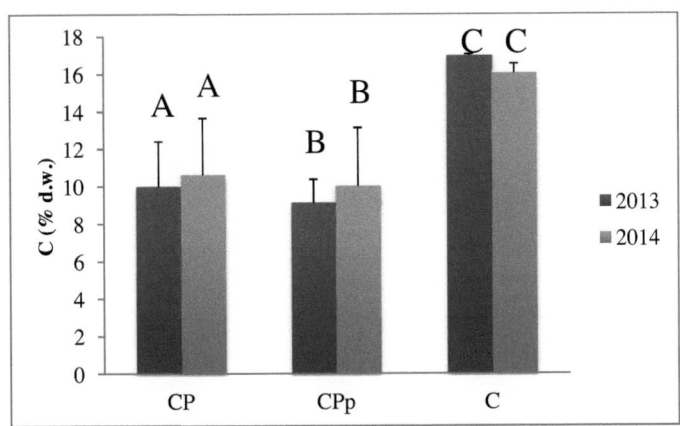

Figure 12. Average (± standard deviation) Carbon content (% d.w.) of substrates with different treatments sampled in autumn 2013 and spring 2014.
CP = Compost + Plants (n = 18); CPp = Compost + Plants + poultry manure (n = 18); C = Compost (n = 3).

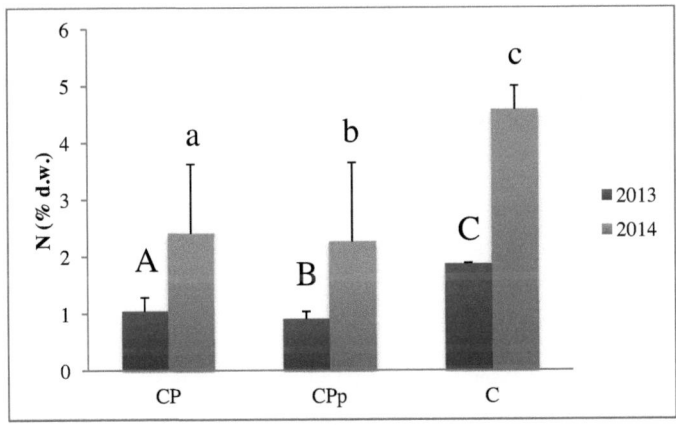

Figure 13. Average (± standard deviation) Nitrogen content (% d.w.) of substrates with different treatments sampled in autumn 2013 and spring 2014.
CP = Compost + Plants (n = 18); CPp = Compost + Plants + poultry manure (n = 18); C = Compost (n = 3).

3.2 *Cd, Cr, Cu, Ni, Pb CONCENTRATIONS IN SAMPLED SUBSTRATES*

Figure 14 shows the Cadmium (Cd), Chromium (Cr), Copper (Cu), Nickel (Ni) and Lead (Pb) in the substrates studied.

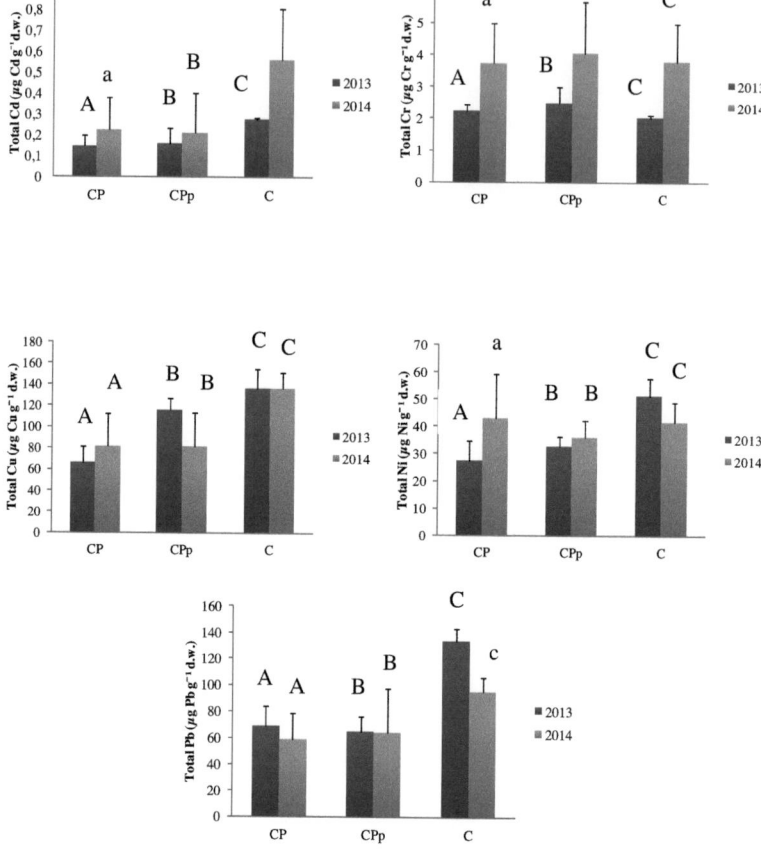

Figure 14. Average (± standard deviation) metal concentrations (μg g^{-1} d.w.) in substrates with different treatments sampled in autumn 2013 and spring 2014.
CP = Compost + Plants (n = 18); CPp = Compost + Plants + poultry manure (n = 18);
C = Compost (n = 3).

We removed obvious *outliers* for total Copper concentrations measured in 2013 in αPFA1, for total Cadmium measured in 2014 in βPFA2 and αPFL3, and for the available Chromium level measured in 2014 in αFA1 and βFL3. These values were at least a factor of 10 higher than all other values measured in these samples.

The total concentrations of investigated metals, in all substrates and for both samples, are within the limit values specified in Annex 2 of Legislative Decree 75/2010 on the subject of organic fertilizers.

Figure 14 shows that the concentrations of Cd, Cr and Ni were higher in the second sampling (March 2014) compared to the first (October 2013). The differences were significant (respectively, $p<0,05$; $p<0,001$; $p<0,001$ and $p<0,001$) for Cd (CP), Cr (CP and CPp) and Ni (CP). Only compost (C) had higher concentrations of Ni in the sample of 2013, although no significant difference occurred.

Cu concentrations were higher in the mesocosms that received the mixture of compost with poultry manure and plants (CPp) in autumn than in spring, although no significant difference occurred. We found a relationship between total Cu concentrations (μg Cu g^{-1} d.w.) and OM content (%), with $R^2 = 0,316$ in 2013 and $R^2 = 0,394$ for the 2014 samples. In fact, Organic Matter has a great ability to complex Cu, probably also explaining for the correlation of Cu levels with organic matter (Martin *et al.*, 2005). The concentrations of Pb were always higher in the first sampling compared to the second, although the differences were significant ($p<0,05$) only for Compost (C).

By contrast with the total concentration, the available metal concentrations were higher in the sampling of 2013 compared to that of 2014 (Figure 15). The differences were significant ($p<0,001$) for Cd (CP and CPp). The differences were significant ($p<0,05$ and $p<0,001$) for Ni (CP and CPp).

However, available Cu levels in the mesocosms amended with the mixture of compost and plants (CP) were higher in spring in autumn, although no significant difference occurred.

Available Pb concentrations in the substrates treated with mixture of compost and plants (CP) were significantly ($p<0,05$) higher in 2014 than in 2013.

The available Cr concentrations were always higher in the second sampling compared to the first (Figure 15). The differences were significant ($p<0,05$ and p<0,01) for CP and CPp respectively. In this case, both the quantity and quality of organic matter in 2014 compared to 2013 are likely responsible for the higher availability of Cr in 2014.

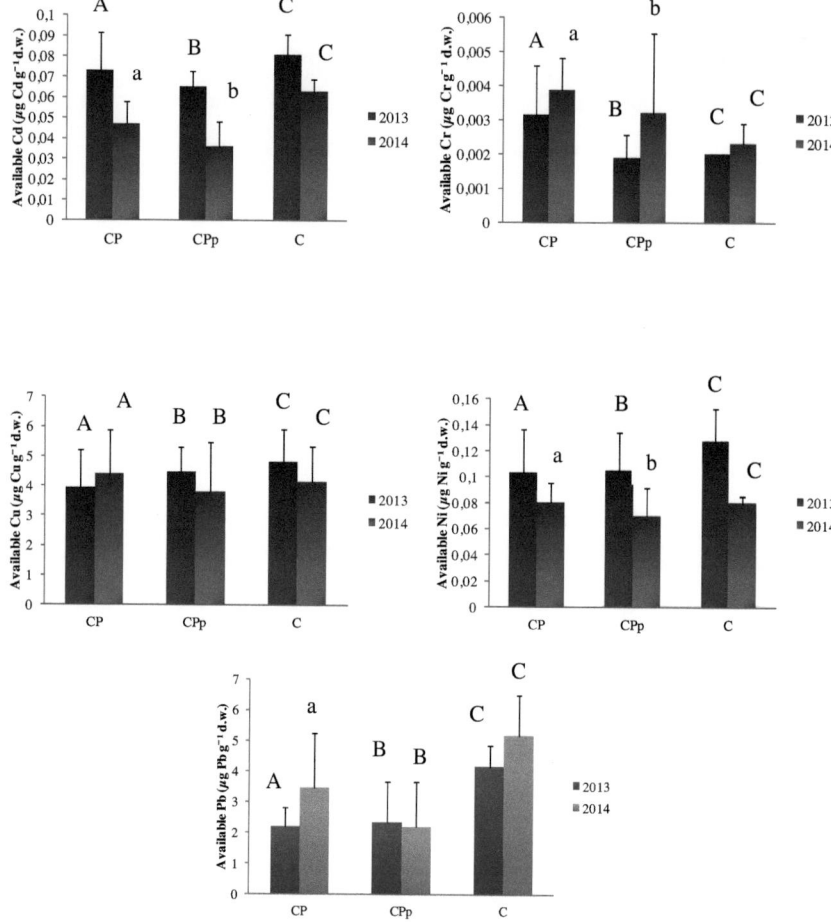

Figure 15. Mean available metal concentrations in substrates with different treatments sampled in autumn 2013 and spring 2014. Error bars indicate standard deviations.
CP = Compost + Plants (n = 18); CPp = Compost + Plants + poultry manure (n = 18);
C = Compost (n = 3).

3.3 BIOLOGICAL CHARACTERIZATION OF MESOCOSMS

For the biological characterization, the following parameters were evaluated: respiration (μg CO_2 g^{-1} d.w. h^{-1}); qCO$_2$ (μg C-CO$_2$/ mg C-mic); CEM (μg C-CO$_2$/ g Corg); TBF (mg g^{-1} d.w.); Cmic (μg C-CO$_2$ g^{-1} d.w.).

We removed obvious *outliers* for respiration (μg CO_2 g^{-1} d.w. h^{-1}), for qCO$_2$ (μg C-CO$_2$/ mg C-mic), and for CEM (μg C-CO$_2$/ g Corg), because these values were negative.

Microbial biomass, Cmic, was significantly ($p<0,001$) higher in autumn 2013 than in spring 2014, but the opposite was true for the metabolic quotient, qCO$_2$ (Figure 16), in fact the differences were significant ($p<0,001$).

The total fungal biomass, TFB, the coefficient of endogenous mineralization, CEM, and respiration showed the same temporal pattern, being highest in 2013 compared to 2014 in substrates treated with the mixture of compost and plants (CP), and lowest in 2013 in the substrates amended with the mixture of compost with plants and poultry manure (CPp) as well as the compost only treatments (C) (Figure 16), although no significant difference occurred.

Microbial activity (CEM) and biomass were higher in CPp treatments compared to C and to CP, although no significant difference occurred. Microbial biomass and activity play an important role in nutrient cycling. Our findings agree with of Bhattacharyya *et al.* (2005) who found higher microbial biomass in submerged rice soils amended with decomposed cow manure than in soils amended with municipal solid waste compost, but our results did not agree regarding the metabolic quotient.

The positive effect of the addition of poultry manure to compost on soil microbial biomass as well as on nutrient availability appeared to decrease over the study period. This was, however, no the case for organic matter content and Cr levels of the soils.

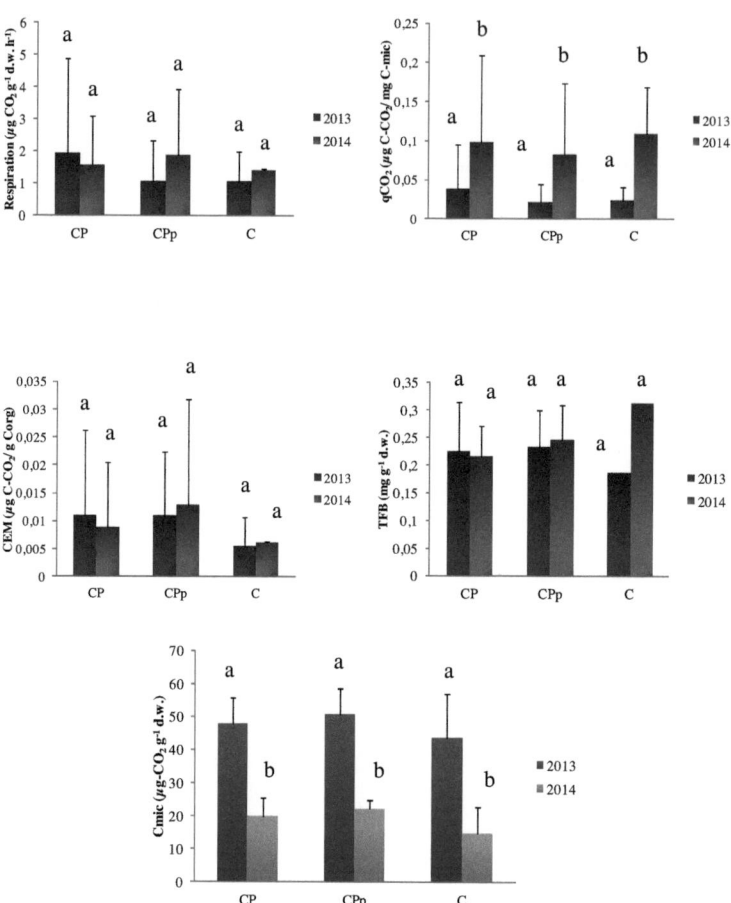

Figure 16. Mean microbial biomass (C_{mic}), total fungal biomass (TFB), respiration, metabolic quotient (qCO_2) and coefficient of endogenous mineralization (CEM) in mesocosms amended with compost plus plants (CP), compost with poultry manure and plants (CPp) and with compost (C). Substrates were incubated in mesocosms and sampled in autumn 2013 and spring 2014.
Error bars indicate standard deviations. For Respiration, qCO_2 and Cmic, n = 18 (CP 2013 and CPp 2013); n = 12 (CP 2014 and CPp 2014); n = 3 (C 2013); n = 2 (C 2014). For CEM, n = 17 (CP 2013); n = 13 (CP 2014 and CPp 2014); n = 16 (CPp 2013); n = 3 (C 2013); n = 2 (C 2014). For TFB, n = 6 (CP and CPp); n = 1 (C).

Soil microbial biomass [Cmic (μg C-CO$_2$ g^{-1} d.w.)] can vary in different climate regimes, and is generally higher in regions with colder and humid climate than in warmer and drier regions. Furthermore, also some chemical and physical properties of the soil are dependent on the climate regime, such as temperature, humidity and pH. And their seasonal fluctuations affect the microbial biomass of the soil. The higher water holding capacity of the mesocosms amended with CPp substrate has a positive feedback on microbial populations enhancing their biomass and activity (Bhattacharyya et al., 2001; Kunito et al., 2001).

Respiration (μg CO_2 g^{-1} d.w. h^{-1}), measured as CO_2 evolution, is an estimate of the metabolism of soil biota; if the edaphic community is richer and more active, the evolution of carbon dioxide will be greater. Generally, soil respiration rate follows the organic matter content. In agreement with the work of Jones et al. (2008), respiration is linearly related to the quantity of organic carbon in the soil.

Anderson and Domsch (1978) proposed an ecophysiological index, derived from the rate of respiration per unit of microbial biomass and time unit, to evaluate the metabolic efficiency of the soil microbial community. Higher values of metabolic quotient [qCO_2 (μg C-CO_2/ mg C-mic)] imply greater microbial activity. In practice, the microorganisms require more carbon and spend more energy on respiration rather than increasing their biomass. In fact, the lower qCO_2 values in CPp as well as in the sampling of 2013 indicate less stressful conditions than in the C treatments as well as in the sampling of 2014 (Anderson and Domsch, 1993).

The coefficient of endogenous mineralization [CEM (μg C-CO_2/ g Corg)] is derived from the ratio of microbial carbon and organic carbon in the soil. Variations of CEM are related to the dynamics of organic matter and may indicate leaks or accumulation of soil carbon in the soil (Anderson et al. 1986). Manure is degraded faster than plant debris and mineralization is improved. This is consistent with the higher coefficient of endogenous mineralization detected in CPp amended substrates (Maisto et al., 2010). No relationship was found between CEM and Organic Matter content (%) in 2013, but a negative trend was observed in 2014, with $R^2 = 0,153$.

Among the indices microbial, a component of the total microbial biomass is the fungal community. Fungi play an important role in nutrient recycling and have been shown to have a different response to disturbances than other indices such as microbial biomass or respiration. In this study, the total fungal biomass [TFB (mg g^{-1}

d.w.)] in the CP treatment was lower in the sampling of 2014 compared to 2013, but it was higher in 2014 for the CPp and C treatments, although no significant difference between substrates occurred. It is interesting to note that compost had a suppressive effect on soil borne fungal plant pathogens, and its application was also effective in plant diseases control (Bonanomi *et al.*, 2007). In addition, Larchevêque *et al.* (2005) showed that compost induces decreasing colonization of leaf litter by fungi but not by bacteria. Moreover, Zaccardelli *et al.* (2006) report that spore-forming bacteria contained in a compost obtained from the organic fraction of municipal solid wastes showed antibiotic activity against three species of fungi. No relationship was found between Total Fungal Biomass (mg g^{-1} d.w.) and Organic Matter content (%) in 2013, but a negative trend was observed in 2014. In this case $R^2 = 0,293$. Moreover, in the literature it is well-known that fungi are sensitive multiple and various edaphic and climatic factors, such as different types of disturbance, type of soil and vegetation, quantity and quality of the organic matter, temperature, precipitation, and use of fertilizers.

3.4ECOTOXICOLOGICAL TESTS

We determined the germination and the root extension of *Lepidium sativum* and *Sorghum saccharatum*, and the survival and growth of *Heterocypris incongruens* exposed to different test samples.

Germination Index takes into account two endpoints: seed germination and root extension. Figure 17 shows that germination (E %) of *L. sativum* was higher on 2013 samples from compost with plants treatments and also higher in 2014 on samples from treatments with compost, plants and poultry manure, although no significant difference occurred. Germination (E %) of *S. saccharatum* was always higher on 2013 samples. The differences were significant (p<0,05 and p<0,01) for CP and CPp. The different responses, in addition to the differences between the species of plants used, could be due to the combination of the physical-chemical properties of the soils with the concentrations of the metals, but there is no relation.

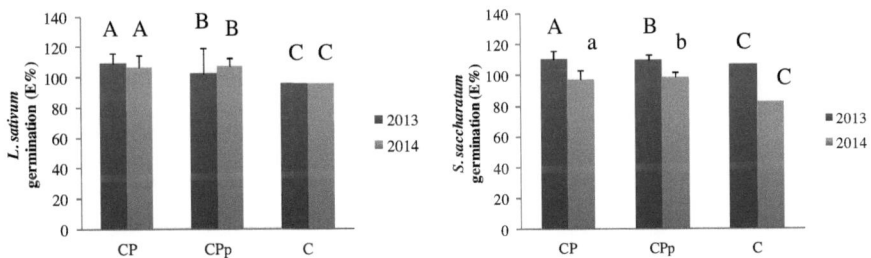

Figure 17. Mean Germination (E%) of *Lepidium sativum* and *Sorghum saccharatum* in substrates with different treatments sampled in autumn 2013 and in spring 2014.
Error bars indicate standard deviations.
CP = Compost + Plants (n = 6); CPp = Compost + Plants + poultry manure (n = 6);
C = Compost (n = 1).

Similarly to studies by other authors (Wong *et al.*, 2001; Fuentes *et al.*, 2006), germination ability proved to be the least sensitive parameter. Most published studies indicate that root extension is a more sensitive parameter than seed germination.

Figure 18 shows that the root extension (E %) was significantly (p<0,001) higher on 2014 samples, for both *L. sativum* and *S. saccharatum* in the CP and CPp treatments.

On the 2014 samples *L. sativum* root extension showed a negative relationship total Cu concentration in the soils (μg Cu g^{-1} d.w.). On the other hand, Arambašic *et al.* (1995) compared the toxicity of Cu, Pb and Zn and phenols to *L. sativum*. This species is found to be sensitive to these compounds in the following order: phenol> Cu> Pb> Zn.

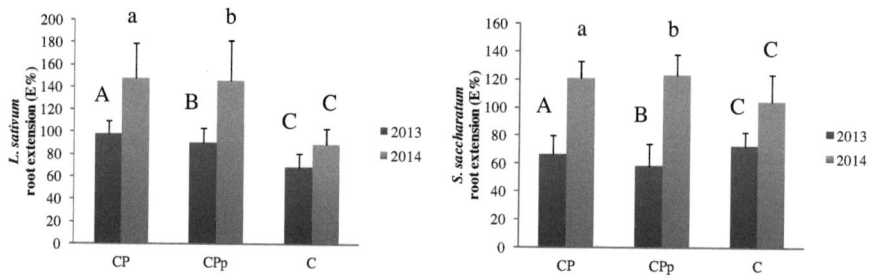

Figure 18. Mean Root Extension (E%) of *Lepidium sativum* and *Sorghum saccharatum* in substrates with different treatments sampled in autumn 2013 and in spring 2014.
Error bars indicate standard deviations.
CP = Compost + Plants (n = 18); CPp = Compost + Plants + poultry manure (n = 18);
C = Compost (n = 3).

Figure 19 shows that the survival of *H. incongruens* was significantly (p<0,01) greatest on CPp samples from the first sampling in autumn 2013 for the mixture of compost with poultry manure and plants (CPp). The growth of *H. incongruens* was always significantly (p<0,001; p<0,001; p<0,05) greatest on substrates from the first sampling.

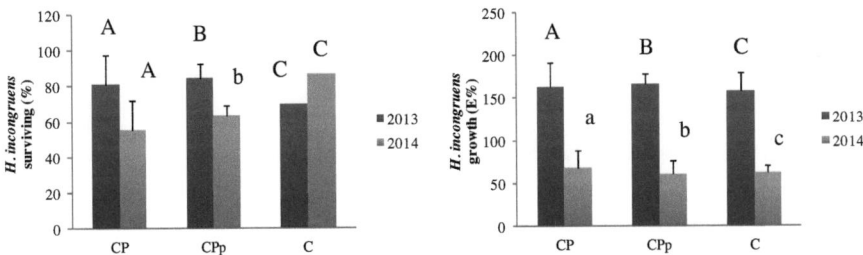

Figure 19. Mean Survival (%) and Growth (E%) of *Heterocypris incongruens* in substrates with different treatments sampled in autumn 2013 and spring 2014. Samples were taken from mesocosms treated with compost plus plants (CP), with compost with poultry manure and plants (CPp) and with compost (C).

Error bars indicate standard deviations.

For Surviving n = 6 (CP and CPp); n = 1 (C). For Growth, n = 18 (CP and CPp); n = 3 (C).

3.5 METAL BIOAVAILABILITY TO THE EARTHWORM EISENIA ANDREI

Earthworms (*Eisenia andrei*) were used as test organisms to assess bioavailability of elements in the test samples. In earthworms, concentrations of 3 metals (Cd, Ni, Pb) were measured.

Figure 20 shows only Cd (*detection limit: 0,003 $\mu g/g$*), because the concentrations for Ni (*detection limit: 0,001 $\mu g/g$*) and Pb (*detection limit: 0,03 $\mu g/g$*) were very low and often below detection limit. This figure shows earthworm concentrations. And the earthworm concentrations seem very low, so that indeed indicates our soils are pretty clean.

Cd is a non-essential metal. Uptake (and also elimination) of Cd in soil organisms is highly species-dependent, although Ardestani *et al.* (2014) did not see a clear phylogenetically-related trend in cadmium uptake rate constants.

No earthworm mortality was found in the substrates and also any statistical difference was found (the analysis of the data was performed by one-way ANOVA).

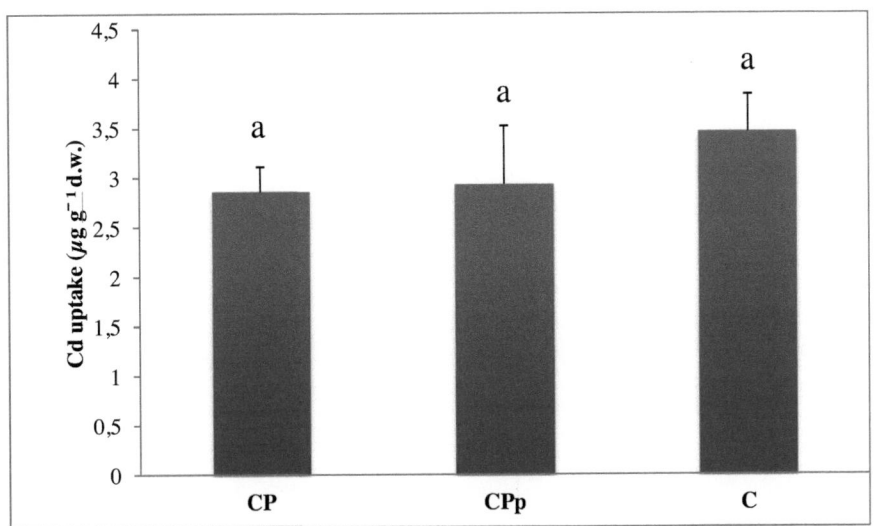

Figure 20. CADMIUM UPTAKE IN EARTHWORMS (*Eisenia andrei*) exposed for 3 weeks in samples with different treatments sampled from mesocosms. Average values of cadmium concentration (μg/g). Error bars indicate standard deviations.
CP = Compost + Plants (n = 3); CPp = Compost + Plants + Poultry manure (n = 3);
C = Compost (n = 5).

Figure 21 shows the weight of earthworms at the beginning and at the end of the experiment, expressed in *percent* (%). All earthworms lost weight, without significant differences depending on treatments of substrates (the STATISTICAL analysis of the data was performed by one-way ANOVA).

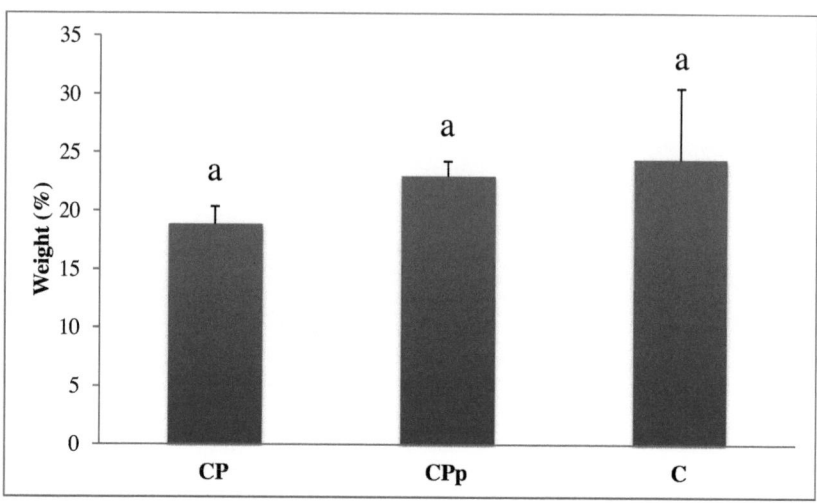

Figure 21. Average weight losses in earthworms (*Eisenia andrei*) exposed for 3 weeks in substrates with different treatments sampled from mesocosms. Average weight change (%). Error bars indicate standard deviations.
CP = Compost + Plants (n = 3); CPp = Compost + Plants + Poultry manure (n = 3); C = Compost (n = 5).

After 21 days, food added to the containers was hardly consumed. The changes in weight may bias the element concentration in earthworms. Weight loss may indirectly result in a relatively higher body concentration of metals or the uptake can be directly affected due to changes in earthworm behaviour. However in this work, in agreement with *García*-Gómez *et al*. (2014), the changes in weight did not appear to affect the observations.

4. 6 STATISTICS

We used a two-tailed Student's t-Test with assumption of equal variances in Excel.

The t-Test was used to test the null hypothesis that the means of two "populations" are equal.

With statistical analysis of results for pH (t-Test 2-tailed), we have seen that:

✓ p<0,01 (0,00024) for CP 2013 *versus* CP 2014;

✓ p<0,01 (0,007645) for CPp 2013 *versus* CPp 2014.

With statistical analysis of results for Organic Matter (t-Test 2-tailed), we have seen that:

✓ p<0,05 (0,026466) for CP 2013 *versus* CPp 2013.

The other significance values are given in the text.

Relationships between physico-chemical characterization of substrates, biological characterization of substrates and ecotoxicological tests

To determine whether it is possible to establish a relationship between the physico-chemical properties of substrates, biological characterization of substrates and ecotoxicological tests, parameters were compared.

Total Cu concentration in the substrates (μg Cu g^{-1} d.w.) was correlated with OM (%) (Figure 22), with $R^2 = 0,316$ (2013 samples) and $R^2 = 0,394$ (2014).

No relationship was found between Coefficient of Endogenous Mineralization (μg C-CO$_2$/ g Corg) and Organic Matter content (%) in 2013 samples, but a negative trend was observed in 2014 samples (Figure 23).

No relationship was found between Total Fungal Biomass (mg g^{-1} d.w.) and Organic Matter content (%) in 2013, but a negative trend was observed in 2014 (Figure 24). In this case $R^2 = 0,293$.

No relationship was found between *L. sativum* root extension (E %) and Total Cu concentrations (μg Cu g^{-1} d.w.) in 2013 samples ($R^2 = 0,111$), but a negative trend was observed in 2014 samples, where $R^2 = 0,205$ (Figure 25).

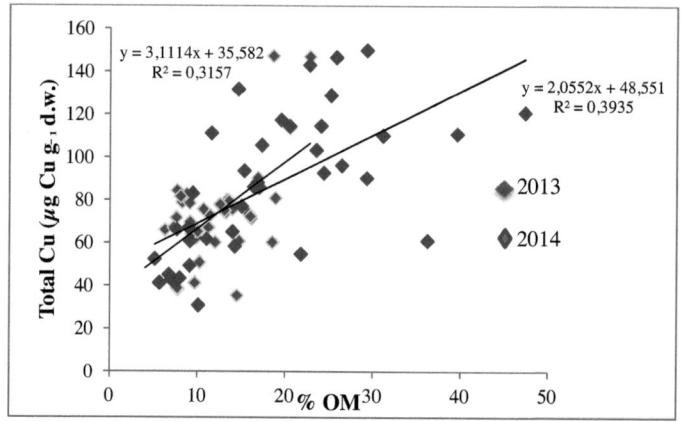

Figure 22. Relationship between Total Cu concentration (μg Cu g^{-1} d.w.) and organic matter content (% OM) in substrates with different treatments sampled in autumn 2013 and spring 2014. Samples were taken from mesocosms treated with compost plus plants (CP), with compost with poultry manure and plants (CPp), and with compost (C).

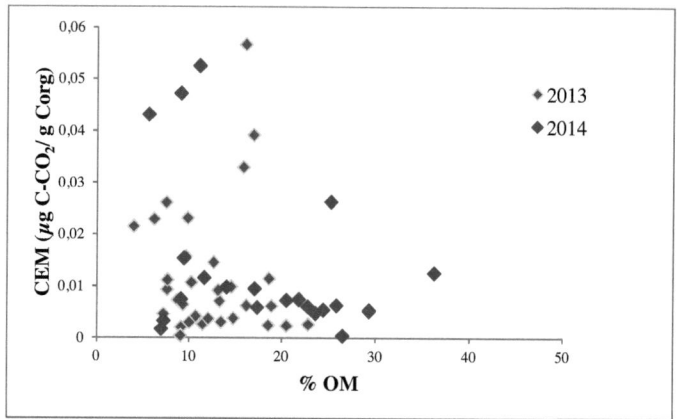

Figure 23. Relationships between Coefficient of Endogenous Mineralization (μg C-CO$_2$/ g Corg) and organic matter content (% OM) in substrates with different treatments sampled in autumn 2013 and spring 2014. Samples were taken from mesocosms treated with compost plus plants (CP), with compost with poultry manure and plants (CPp), and with compost (C).

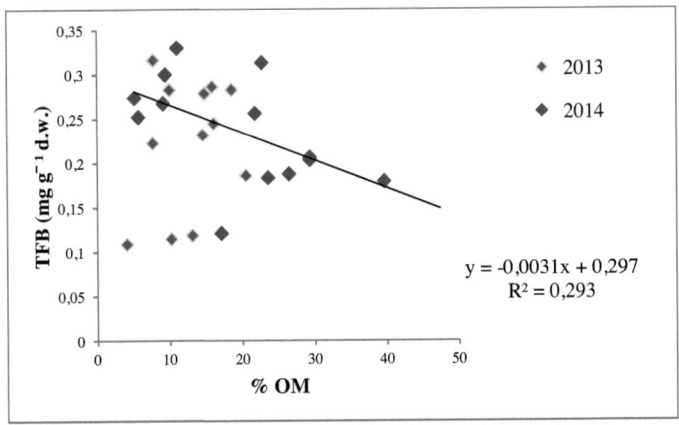

Figure 24. Relationships between Total Fungal Biomass (TFB; mg g^{-1} d.w.) and organic matter content (% OM) in substrates with different treatments sampled in autumn 2013 and spring 2014. Samples were taken from mesocosms treated with compost plus plants (CP), with compost with poultry manure and plants (CPp), and with compost (C).

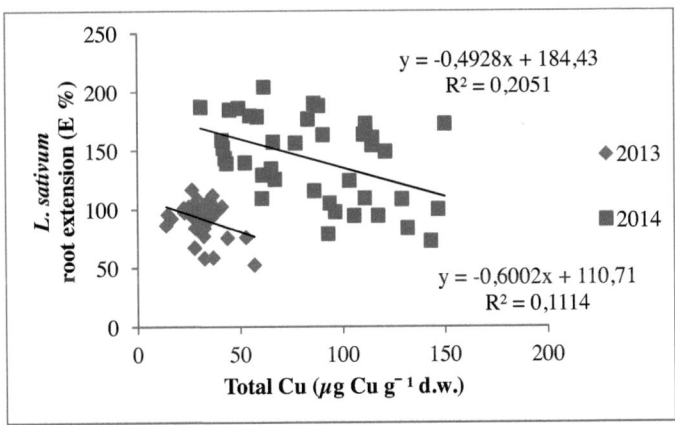

Figure 25. Relationships between *L. sativum* root extension (E %) and Total Cu concentration (μg Cu g^{-1} d.w.) in substrates with different treatments sampled in autumn 2013 and spring 2014. Samples were taken from mesocosms treated with compost plus plants (CP), with compost with poultry manure and plants (CPp), and with compost (C).

4. CONCLUSION (CHAPTER 5)

The **pH** of the investigated substrates in withdrawals in autumn was around neutrality (around 7), while the pH of the substrates sampled in spring was slightly basic (around 8). The only-compost-amended substrates (C) at both sampling times had the highest **organic matter** content. The **available metal concentration** was higher in the autumn 2013 samples (for Cd and Ni) compared to spring 2014; however the available Pb concentrations in the substrates treated with mixture of compost and plants (CP) were higher in 2014 than in 2013. **Microbial biomass** in the mesocosms was higher at the first sampling, whereas **metabolic quotient** was higher at the second. Also, microbial biomass and activity were higher in substrates amended with the mixture of compost and poultry manure than in the substrates that received only compost. The results of this study demonstrate that the use of organic compost as a substrate for sclerophyllous shrubs is feasible and may be helpful in mitigating the environmental impact of organic wastes disposed on the soil.

In the phytotoxicity evaluation, **germination** and **root extension** for two plants (*L. sativum* and *S. saccharatum*) were determined. *S. saccharatum* showed always higher germination on 2013 samples. Root extension (E %) was higher on 2014 samples, both for *L. sativum* and for *S. saccharatum*. *H. incongruens* is considered particularly suitable for toxicity tests of the soil because it is sensitive to organic and inorganic pollutants. **Survival** and **growth** of *H. incongruens* was highest in substrates from the first sampling.

To assess metal bioavailability, earthworms *Eisenia andrei* were exposed to the different substrates. No earthworm mortality was found in the test samples. *Ergo*, our soils were not contaminated.

- *The data presented suggest that Compost with and without additions of plant material and poultry manure does affect available metal concentrations and effect was observed on plants (*L. sativum *and* S. saccharatum*)*, Heterocypris incongruens *and* Eisenia andrei.

- *The data presented suggest that it is without risk to apply **compost** directly on limestone debris with sclerophyllous shrubs.*

REFERENCES (CHAPTER 6)

✓ Albiach, R., Canet, R., Pomares, F., Ingelmo, F., 2000. *Microbial biomass content and enzymatic activities after the application of organic amendments to a horticultural soil.* Bioresource Technology 75: 43-48.

✓ An, Y.-J., 2004. *Soil ecotoxicity assessment using cadmium sensitive plants.* Environmental Pollution 127: 21-26.

✓ Anderson, J. P. E., Domsch, K. H., 1978. *A physiological method for the quantitative measurement of microbial biomass in soil.* Soil Biology & Biochemistry 10: 215-221.

✓ Anderson, T. H., Domsch, K. H., 1993. *The metabolic quotient for CO_2 (qCO_2) as a specific activity parameter to assess the effects of environmental conditions, such as pH, on the microbial biomass of forest soils.* Soil Biology & Biochemistry 25: 393-395.

✓ APAT, 2004. *Guida tecnica su metodi di analisi per il suolo e i siti contaminati: utilizzo di indicatori biologici ed ecotossicologici.* RTI CTN-TES 1/2004.

✓ APAT, 2006. *L'ecotossicologia negli ambienti acquatici.* Rapporto 71/2006.

✓ Arambašic, M. B., Bjeli, S., Subakov, G., 1995. *Acute toxicity of heavy metals (copper, lead, zinc), phenol end sodium on* Allium cepa *L.,* Lepidium sativum *L. and* Daphnia magna *St.: comparative investigations and the practical applications.* Water Research 29(2): 497-503.

✓ Ardestani, M. M., Van Straalen, N. M., Van Gestel, C. A. M., 2014. *Uptake and elimination kinetics of metals in soil invertebrates: A review.* Environmental Pollution 193: 277-295.

✓ Baldantoni, D., Leone, A., Iovieno, P., Morra, L., Zaccardelli, M., Alfani, A., 2010. *Total and available soil trace element concentrations in two*

Mediterranean agricultural systems treated with municipal waste compost or conventional mineral fertilizers. Chemosphere 80: 1006-1013.

✓ Barzegar, A. R., Yousefi, A., Daryashenas, A., 2002. *The effects of addition of different amounts and types of organic materials on soil physical properties and yield of wheat.* Plant Soil 247: 295-301.

✓ Bhattacharyya, P., Pal, R., Chakraborty, A., Chakrabarti, K., 2001. *Microbial biomass and activity in a laterite soil amended with municipal solid waste compost.* Journal of Agronomy and Crop Science 187: 207-211.

✓ Bhattacharyya, P., Chakrabarti, K., Chakraborty, A., 2005. *Microbial biomass and enzyme activities in submerged rice soil amended with municipal solid waste compost and decomposed cow manure.* Chemosphere 60: 310-318.

✓ Berg, B., Söderström, B. 1979. *Fungal biomass and nitrogen in decomposing Scots pine needle litter.* Soil Biology & Biochemistry 11: 339-341.

✓ Bonanomi, G., Antignani, V., Pane, C., Scala, F., 2007. *Suppression of soil borne fungal diseases with organic amendments.* Journal of Plant Pathology 89: 311-324.

✓ Bullini, L., Pignatti, S., Virzo De Santo, A., 1998. *Ecologia generale.* UTET (Ed.), Torino, pp. 519.

✓ Butler, G. C., 1978. *Principles of Ecotoxicology*, SCOPE 12. Wiley and Sons (Ed.), Chichester, England, pp. 350.

✓ Caravaca, F., Garcia, C., Hernandez, M. T., Roldan, A., 2002. *Aggregate stability changes after organic amendment and mycorrhizal inoculation in the afforestation of a semiarid site with Pinus halepensis.* Applied Soil Ecology 19: 199-208.

✓ Centemero, M., Caimi, V., 2001. *Impieghi del compost: settori di maggior rilevanza, modalità d'uso, scenari attuali di mercato.* Acts Course Compost: production and use, Ed. CIC Rimini.

✓ Centemero, M., Zanardi, W., 2006. *Il compostaggio nella gestione dei rifiuti urbani biodegradabili*. Day-depth technical-scientific: Biological processes for the management of municipal waste.

✓ Certini, G., Ugolini F. C., 2013. *An updated, expanded, universal definition of soil*. Geoderma 192: 378-379.

✓ Chial, B., Persoone, G., 2003. *Cyst-based toxicity tests XV-Application of ostracod solid-phase microbiotest for toxicity monitoring of contaminated soils*. Environmental Toxicology 18(5): 347-352.

✓ Chung, M. K., Hu, R., Wong, M. H., Cheung, K. C., 2007. *Comparative toxicity of hydrophobic contaminants to microalgae and higher plants*. Ecotoxicology 16: 393-402.

✓ Ciccotti, A. M., De Clauser, R., Zorzi, G., Cristoforetti, A., Silvestri, S., Jodice, R., Gasperi, F., Pinamonti, F., 1988. *Trasformazione in compost di residui di natura diversa. Aspetti biochimici e microbiologici nel corso del processo*. Experiences and research XVIII: 75-120.

✓ Crecchio, C., Curci, M., Pizzigallo, M. D. R., Ricciuti, P., Ruggiero, P., 2004. *Effect of municipal soil waste compost amendments on soil enzyme activities and bacterial genetic diversity*. Soil Biology & Biochemistry 36: 1595-1605.

✓ Crouau, Y., Pinelli, E., 2008. *Comparative ecotoxicity of three polluted industrial soils for the Collembola* Folsomia candida. Ecotoxicology and Environmental Safety 71: 643-649.

✓ Dokuchaev, V. V., 1880. *Protocol of the meeting of the branch of geology and mineralogy of the Petersburg Society of Naturalists*. [Translated by the Department of Soils and Plant Nutrition, University California, Berkeley]. Transactions of the St. Petersburg Society of Naturalists XII: 65-97.

✓ EPA, 1996. *Ecological effects test guidelines*. OPPTS 850.4200. Seed Germination/Root Elongation Toxicity Test. EPA-712-C-96-154.

✓ European Commission's Directive on Soil Strategy. *The implementation of the Soil Thematic Strategy and ongoing activities*, COM(2012) 46 (13 February 2012).

✓ Federer, C. A., Hornbeck, J. W., 1985. *The buffer capacity of forest soils in New England*. Water Air Soil Pollution 26: 163-173.

✓ Fischer, A. G., 1960. *Latitudinal variation in organic diversity*. Evolution 14: 64-81.

✓ Fuentes, A., Llorens, M., Sacz, J., Aguilar, M. I., Perez-Marin, A. B., Ortuno, J. F., Meseguer, V. F., 2006. *Ecotoxicity, phytotoxicity and extractability of heavy metals from different stabilized sewage sludges*. Environmental Pollution 143: 355-360.

✓ Gallardo-Lara, F., Nogales, R., 1987. *Effect of the application of town refuse compost on the soil-plant system: a review*. Biological Wastes 19: 35-62.

✓ García-Gómez, C., Sánchez-Pardo, B., Esteban, E., Peñalosa, J. M., Fernández, M. D., 2014. *Risk assessment of an abandoned pyrite mine in Spain based on direct toxicity assays*. Science of the Total Environment 470-471: 390-399.

✓ Hamdi, H., Benzarti, S., Manusadžianas, L., Aoyama, I., Jedidi, N., 2007. *Solid-phase bioassays and soil microbial activities to evaluate PAH-spiked soil ecotoxicity after a long-term bioremediation process simulating landfarming*. Chemosphere 70: 135-143.

✓ Hubálek, T., Vosáhlová, S., Mateju, V., Kovácová, N., Novotný, C., 2007. *Ecotoxicity monitoring of hydrocarbon-contaminated soil during bioremediation: A case study*. Archives of Environmental Contamination and Toxicology 52: 1-7.

✓ Iglesias-Jimenez, E., Alvarez, C., 1993. *Apparent availability of nitrogen in composted municipal refuse*. Biology and Fertility of Soils 16: 313-318.

✓ Ihnat, M., Fernandes, L., 1996. *Trace elemental characterization of composted poultry manure*. Bioresource Technology 57: 143-56.

✓ Jenkinson, D. S. and Ladd, J. N., 1981. *Microbial biomass in soil: measurements and turnover*. In Soil biochemistry (Paul, E. A., Ladd, J. N., eds) Inc. New York and Basel 415-471.

✓ Jenny, H., 1980. *The soil resource: origin and behavior*. Ecology Studies 37. Springer-Verlag, New York, NY, USA.

✓ Jones, D. L., Brassington, D. S., 1998. *Sorption of organic acids in acid soils and its implications in the rhizosphere*. European Journal of Soil Science 49: 447-455.

✓ Kaiser, K. and Zech, W., 2000. *Sorption of dissolved organic nitrogen by acid subsoil horizons and individual mineral phases*. European Journal of Soil Science 51: 403-411.

✓ Kunito, T., Saeki, K., Goto, S., Hayashi, H., Oyaizu, H., Matsumoto, S., 2001. *Copper and zinc fractions affecting microorganisms in long-term sludge-amended soils*. Bioresource Technology 79: 135-146.

✓ Larchevêque, M., Baldy, V., Korboulewsky, N., Ormeño, E., Fernandez, C., 2005. *Compost effect on bacterial and fungal colonization of kermes oak leaf litter in a terrestrial Mediterranean eco system*. Applied Soil Ecology 30: 79-89.

✓ Leitgib, L., Kálmán, J., Gruiz, K., 2007. *Comparison of bioassays by testing whole soil and their water extract from contaminated sites*. Chemosphere 66: 428-434.

✓ Lindsay, W. L., Norwell, W. A., 1978. *Development of a DTPA soil test for zinc, iron, manganese and copper*. Soil Science Society of American Journal 42: 421-428.

✓ Maffiotti, A., Bona, F., Volterra, L., 1997. *Introduzione all'ecotossicologia*. Analisi e recupero dei sedimenti marini. Quaderni di Tecniche di Protezione Ambientale. Pitagora (Ed.), Bologna, pp. 139.

✓ Maisto, G., De Marco, A., De Nicola, F., Arena, C., Vitale, L., Virzo De Santo, A., 2010. *Suitability of two types of organic wastes for the growth of sclerophyllous shrubs on limestone debris: A mesocosm trial*. Science of the Total Environment 408: 1508-1514.

✓ Maisto, G., 2013. Ecotoxicology. *Lecture in the course Ecotoxicology*, November 20.

✓ Martin, M., Menardo, S., Barberis, E., Brugiafreddo, M., Bourlot, G., 2005. *Fitotossicità da rame in suoli acidi ex-vitati*. Quaderni Regione Piemonte - Agricoltura 42: 35-39.

✓ Nahmani, J., Hodson, M. E., Black, S., 2007. *A review of studies performed to assess metal uptake by earthworms*. Environmental Pollution 145: 402-424.

✓ Odum, E. P., Barrett, G. W., Richard, B., 2004. *Fundamentals of Ecology*. Cengage Learning.

✓ OECD, 1984. *Terrestrial plants: growth test*. OECD Guidelines for Testing of Chemicals 208, Organization for Economic Cooperation and Development, Paris, France.

✓ Olson, F. C. W. 1950. *Quantitative estimates of filamentous algae*. Transaction of the American Microscopy Society 69: 272-279.

✓ Reich, J. W., Schlesinger, W. H., 1992. *The global carbon dioxide flux in soil respiration and its relationship to climate*. Tellus 44 B: 81-99.

✓ Roh, J.-Y., Sim, S. J., Yi, J., Park, K., Chung, K. H., Ryu, D.-Y., Choi, J., 2009. *Ecotoxicity of silver nanoparticles on the soil nematode* Caenorhabditis elegans *using functional ecotoxicogenomics*. Environmental Science & Technology 43: 3933-3940.

✓ Scagliarini, S., 1999. *Le funzioni del compost nel terreno*. Il Divulgatore Anno XXII. Centro Agricoltura Ambiente.

✓ Smith, T. M., Smith, R. L., 2009. *Elementi di Ecologia*. PEARSON-Benjamin Cummings.

✓ Soumarè, M., Tack, F. M. G., Verloo, M. G., 2003. *Characterisation of Malian and Belgian solid waste composts with respect to fertility and suitability for land application*. Waste Management 23: 517-522.

✓ Tarradellas, J., Bitton, G., Rossel, D., 1996. *Soil Ecotoxicology*. (Ed.), CRC Press, Boca Raton, pp. 208-210.

✓ United Nations Convention to Combat Desertification in Those Countries Experiencing Serious Drought and/or Desertification, Particularly in Africa (*UNCCD*), Paris, 14 October 1994.

✓ Van-Camp, L., Bujarrabal, B., Gentile A-R.; Jones R. J. A., Montanarella, L., Olazabal, C., Selvaradjou S-K., 2004. *Reports of the Technical Working Groups Established under the Thematic Strategy for Soil Protection.* EUR 21319 EN/3, Office for Official Publications of the European Communities, Luxembourg.

✓ Van Eck, H., Stewart, B. A., 1995. *Manure.* In: Rechcigl, J. E., editor. Soil amendments and environmental quality. Boca Raton, Florida: CRC Publishers Inc.: 169-198.

✓ Van Gestel, C. A. M., 2012. *Soil ecotoxicology: state of the art and future directions.* ZooKeys 176: 275-296.

✓ Van Hees, P. A. W., Vinogradoff, S. I., Edwards, A. C., Godbold, D. L., Jones, D. L., 2003. *Low molecular weight organic acid adsorption in forest soils: effects on soil solution concentrations and biodegradation rates.* Soil Biology and Biochemistry 35: 1025-1026.

✓ Walker, C. H., Sibly, R. M., Hopkin, S. P., Peakall, D. B., 2012. *Principles of Ecotoxicology.* CRC Press, Boca Raton, pp. 313-322.

✓ Wong, J. W. C., Li, K., Su, M., Fang, D. C., 2001. *Toxicity evaluation of sewage sludge in Hong Kong.* Environment International 27: 373-380.

✓ Wright, D. A., Welbourn, P., 2002. *Environmental Toxicology.* Cambridge University Press, Cambridge, U.K., pp. 630.

✓ Zaccardelli, M., Malzone, A., Campanile, F., De Nicola, F., 2006. *Characterization for antibioses activity of spore-forming bacteria isolated from compost.* Journal of Plant Pathology 88: 62.

✓ Zdruli, P., Jones, R. J. A., Montanarella, L., 2004. *Organic Matter in the Soils of Southern Europe*. European Soil Bureau Technical Report, EUR 21083 EN: 16. Office for Official Publications of the European Communities, Luxembourg.

RINGRAZIAMENTI

ACKNOWLEDGMENTS

The author is very grateful to reviewers for their invaluable and thoughtful comments.

La mia Tesi di Laurea triennale è uno scrigno di ricordi.

. . . pensieri profondi, rivolti alle persone fondamentali della mia esistenza.

Quindi, caro Lettore, se vuoi rileggere quelle parole, va' e sfoglia le pagine finali, quelle dei "Ringraziamenti".

È passato solo un anno dalla Laurea ed i pilastri della mia vita non sono cambiati in questo arco di tempo.

Per tale ragione, non scriverò nulla, ma lascerò parlare le immagini, aggiungendo solo qualche didascalia.

Anzi, in questo anno una cosa è variata, mutando anche me stessa: il Progetto ERASMUS.

Un caleidoscopio di esperienze, conoscenze e novità.

Ora, prendi solo il tempo di guardare questi fogli conclusivi e riempi il tuo cuore di gioia!

Ciao,
famiglia!

«Parto. Non dimenticherò né via Toledo né tutti gli altri quartieri di **Napoli**;

ai miei occhi è, senza nessun paragone, la città più bella dell'Universo»

QUESTO E' IL MIO ERASMUS

Una delle esperienze più intense della mia vita.

Essendo me stessa, vivendo.

Imparando una nuova lingua ed una nuova cultura.

In una città incredibile ho trovato le persone perfette.

Abbiamo percorso insieme il Paese, crescendo insieme.

Senza nessun tipo di paura, godendo di ogni momento.

Sperimentando ogni emozione in un'avventura di sei mesi che mi ha aiutato a rivalutare il mio modo di vivere.

Mi sono aperta a nuove cose ed a nuove persone. . . posso solo dire, GRAZIE!

THANKS TO *KEES*!

To everyone, with all my love,

Sinterklaas - Christmas 2014

<u>Amsterdam</u>

Micaela

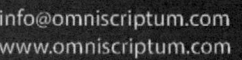

Printed by Books on Demand GmbH, Norderstedt / Germany